A Beautifully Burned Forest

A Beautifully Burned Forest

Learning to Celebrate Severe Forest Fire

Richard L. Hutto

Richard L. Hutto (iD)
Division of Biological Sciences
University of Montana
Missoula, MT, USA

ISBN 978-3-032-03179-2   ISBN 978-3-032-03180-8   (eBook)
https://doi.org/10.1007/978-3-032-03180-8

Cover image: One of the most fire-dependent bird species in the world, the Black-backed Woodpecker, perches on a severely burned Douglas-fir tree in Montana, USA. Photo by Richard L. Hutto

This Springer imprint is published by the registered company Springer Nature Switzerland AG
The registered company address is: Gewerbestrasse 11, 6330 Cham, Switzerland

If disposing of this product, please recycle the paper.

Preface

It was a deadly combination of wind and fire. The winds were so strong that the 2025 Palisades and Eaton fires left little remaining in their wake. The fires were fueled by Santa Ana winds that swept down the San Gabriel Mountains into the Los Angeles basin. They ended up burning through a combined area of about 15,000 ha in just a few days, consuming some 16,000 homes and other structures in the process. A friend from the area texted "unbelievable destruction," and he hit that nail on the head. Had the fire burned through 15,000 ha of mixed-conifer forest without a single home loss in, say, the Sierras or northern Rocky Mountains, a completely different sentiment would have been appropriate, and a celebration would have been in order. Unfortunately, most people do not differentiate between wildland fire and structural fire even though, ecologically speaking, wildland fires are nearly always beautifully restorative events. Wildland fires ought to be viewed completely differently from the way we view fires that sometimes destroy whole communities. I have written this book to explain why this is so.

I am an avian ecologist who likes learning about how birds fit into the larger ecosystem. Early on in my academic career I focused my research attention on deciphering which habitats migratory landbirds use in west Mexico during winter. Later, after the Yellowstone fires of 1988 garnered national attention and created an amazing opportunity to learn more about fire effects, my research direction pivoted to burned forests. Both areas of research were (and still are) politically charged topics. With respect to migratory landbirds, published reports in the late 1980s revealed marked population declines, and scientists suspected that tropical deforestation was responsible for those declines. My own research results showed that migrants were very abundant in second-growth vegetation, which led me to suggest that maybe the primary cause for the reported declines had more to do with land management practices in the United States (e.g., habitat loss and fragmentation) than with deforestation in Mexico. With respect to fire, the ecological patterns I found in burned forests led me to believe that the public at large has been, and is continuing to be, misled by messaging from Smokey Bear and elsewhere. Trying to provide nuance to ever-present public messages that our "forests are out of whack" and that there are "good fires and bad fires" in our forests is not easy, but I persist because I feel confident that these kinds of messages are misleading and not ecologically informed.

Despite the explosion of interest in forest fires these days, there is still surprisingly little ecology associated with the fire science literature. Instead, fire scientists are focused on understanding fire behavior so we can better control that behavior, primarily through timber harvesting and prescribed burning. Simply put, fire scientists undervalue ecology in their studies of fire. This lopsided emphasis in fire science (one largely devoid of ecology) reflects a lack of demand for ecological effects analyses from the public and from professionals who work in the areas of forest and fire management. Accordingly, our public lands are managed with a strong eye toward controlling or subduing the threat of severe fire, but with only a weak eye toward meeting various legal mandates to maintain the ecological integrity of the lands being managed. I hope to make clear how one bird species indicates that severely burned forest conditions are not only highly suitable but also a necessity for it and for many other fire-dependent organisms. I also hope to make clear that we may be falling short when it comes to the ecological part (what ought to be the *main* part) of forest management in relation to those special forest conditions. We ought to value life associated with recently burned conditions as much as we value life associated with long-unburned old-growth conditions. I hope this book will leave you feeling so.

Missoula, MT, USA   Richard L. Hutto

Competing Interests The author has no competing interests to declare that are relevant to the content of this manuscript.

Acknowledgments

Thanks to my wife, Susie, who encouraged me to pull my thoughts on fire ecology together into book form so that whatever biological insight I might have could be shared with others. She also read and re-read various sections and endured years of tweaks until the story became much more readable than what I had originally pulled together from old lecture notes. Rick Halsey was the first person with whom I shared a book draft, and his over-the-top response is what really encouraged me to plow ahead. Bill Baker, Dominick DellaSala, Chad Hanson, and George Wuerthner also read drafts and offered kind words of encouragement to forge ahead at the early stages. Fred Allendorf clarified my thinking surrounding adaptations to fire regimes. I owe special thanks to Sneed Collard, who told me after years of work, to throw out what I had and to start over with a first-person narrative approach; this book would not be nearly as readable as it currently is without his suggestion.

Finally, I am grateful to have received a wide range of research support, which allowed me to gather the data needed to gain a better understanding of the role of fire disturbance in western conifer-forest systems. These sources included the National Geographic Society, USFS Northern

Region, USFS Rocky Mountain Research Station, Glacier National Park, USDI Bureau of Land Management, USDA Cooperative State Research, Education and Extension Service, Montana Fish, Wildlife and Parks, Joint Fire Sciences Program, University of Wyoming/National Park Service Research Center, Teton Science School, Salish-Kootenai Tribes, North American Bluebird Society, Plumb Creek Timber Company, Potlatch Corporation, Champion Timber Corporation, and the University of Montana.

Contents

About the Author

Richard L. Hutto is Professor Emeritus in Biological Sciences and Wildlife Biology at the University of Montana. After joining the faculty in 1977, he taught courses in animal ecology, fire ecology, Montana wildlife, and ornithology across a nearly 40-year career. His early research dealt primarily with the ecology of migratory landbirds throughout the West—in Mexico in winter, the Southwest during spring and fall, and the Northern Rockies in summer. In 1990, he developed the USFS Northern Region Landbird Monitoring Program to generate data on bird distribution patterns so that we might better understand the ecological effects of various land-use practices. To promote informed decision by using data those bird data, Hutto also established the Avian Science Center on the University of Montana campus in 2004. Following the Yellowstone fires of 1988, his research focus shifted toward the ecology of birds in burned forests—an interest he maintains to this day. Dr. Hutto also hosted a nationally televised PBS series called "Birdwatch," which ran from 1998 to 2001.

1

Introduction

Once in my early childhood, mom drove me and my sister through a forested area in the San Gabriel Mountains in our 1952 Oldsmobile. I always liked car rides because the views and the wind through my open window were soothing. I especially liked how mom's wedding ring gave off a sharp ping as it hit the steering wheel every time she maneuvered the car through a bend in the Angeles Crest Highway. Rounding one of those bends, we suddenly came upon a patch of severely burned forest whereupon my sister exclaimed, "Wow, there was a forest fire here!" The only thing I remember mom saying was, "Oh my, kids! That's sad. It's so sad." We all sat in silence passing through what we collectively believed was a destroyed forest. It was not a blackened forest that had just received the *gift* of severe fire; it was a forest that had been *destroyed* by severe fire.

My mom's reaction to the view outside our car was not unexpected, and that's because most people lack the kind of perspective one can achieve only through time spent in a

R. L. Hutto, *A Beautifully Burned Forest*, https://doi.org/10.1007/978-3-032-03180-8_1

burned-forest environment. Extended time with someone or something can change everything about one's perception of that someone or something. The deeper you look, the more you learn. Poet Mary Oliver captured this sentiment well when she wrote that her outdoor work involved "…mostly standing still and learning to be astonished" [1]. But how would anyone be drawn to spend quality time in a burned forest? It usually takes some kind of hook to draw one into spending time in a place that most people find unappealing. Birds, it turns out, are a perfect hook for this sort of thing. No other group of organisms can attract one into a new environment the way birds can. Birdwatchers are eager to visit everything from garbage dumps to pristine forests because different bird species can be found in different environments. Birds are precisely the hook that got me to first wander through a burned forest, so I should begin this story with why I was drawn to use birds as objects of field science.

I was initially attracted to birds for a practical reason. Martin Cody, the professor I wanted to work with for my PhD, used birds as his primary research tool and I wanted to play to his strength. I was in my final year of earning an MS degree when I visited Cody at UCLA to broach the idea of doing PhD work under his tutelage. Having just started watching birds, I remember being excited that I had identified a Yellow-rumped Warbler in some patch of chaparral before our meeting. When I tried to impress him with my ability with bird identification, he merely responded with, "Yeah…it's the most common bird out there right now." It was his way of saying "You got a long way to go with the field end of things, buddy, but you're on the right track."

Birds, I would argue, are the best organisms one can use to monitor environmental effects. This is because birds occur everywhere and are and the most visually and aurally conspicuous vertebrates on the planet. Thus, they are much easier to survey than members of any other living animal group. Birds not only alert you to their presence, but they also identify themselves down to the species (and, sometimes, to the individual) through their conspicuousness, unique behaviors, and unique vocalizations. No need for traps or nets. With nothing more than a good ear, you can record the presence of every bird within a football field's distance around you in a matter of minutes. Name another group of organisms for which this is true. You can't. And because different bird species occur in different environments, they are excellent "indicators" of environmental conditions; a small change in the environment will be accompanied by a change in bird species detected [2]. Birds are perfect tools for indicating the effects of humans on the environment, which is why I chose to use birds as a focus throughout my professional academic career as an ecologist.

Following that childhood trip with my mom and sister, I'm certain I witnessed from afar plenty of burned forests over the ensuing 20 years, but I never took the time to saunter through one until I drove to a 500-ha area of forest that burned in Pattee Canyon just outside Missoula in 1977. By then I had an already established an interest in birds and had been immersed in the study of field ecology for a half-dozen years, and that changed everything. I remember my world view being turned upside down after taking a walk through that burned forest. The burn occurred at the southeastern edge of Missoula in July, right before I moved there from California to take a faculty position in Zoology and Wildlife Biology at the University of Montana. Someone apparently started the fire in the grassland at the bottom of the canyon. Windy conditions then carried flames rapidly into the woods, where they consumed homes and blackened the hillsides. When I drove into that

Fig. 1.1 Two American Three-toed Woodpeckers on a recently burned tree. This uncommon species suddenly became common in Pattee Canyon after the fire of 1977 created ideal conditions for beetle larvae and their woodpecker predators

burn the following spring and got out of the car to explore, there was something refreshing and new about the openness of that freshly burned forest of standing dead trees. I saw birds I never expected to see in a burned conifer forest, like Mountain Bluebirds and American Three-toed Woodpeckers (Fig. 1.1). Some of the plants were striking but unfamiliar to me, like Bicknell's geranium and (although not technically a plant) the fire morel mushroom (Fig. 1.2). I now know that the bluebird is an early burned-forest colonist, the fire morel occurs almost exclusively in the most severely burned portions of a fire, and the geranium can wait for up to 200 years as a seed before severe fire finally comes along and stimulates it to germinate!

By the second winter following the fire, I could enjoy not only the openness, views, and unique plants that the burned area afforded, but also sightings of even more bird species

Fig. 1.2 The fire morel mushroom (*Morchella* sp.) (above) and Bicknell's geranium (*Geranium bicknellii*) (below) are species that rely on severe fire to create the conditions needed for their very existence on planet Earth

and bird behaviors that I had not seen before—Black-backed Woodpeckers foraging in groups with other woodpecker species, and Red Crossbills and Clark's Nutcrackers harvesting seeds from the still-attached cones on Douglas-fir and ponderosa pine trees. Those cones were newly open and exposed on the blackened trees that were still standing. As an ecologist trained to study patterns in nature, it all began to click after multiple visits to that burned forest during the ensuing years.

I love to discover patterns in nature, and what I uncovered in burned forests completely surprised me. If you happen to be a bird watcher, you would already appreciate that different forest conditions support different combinations of bird species, but even bird watchers may not realize that burned forests take this phenomenon to a new level. Indeed, some of the species appear to be unique to, or hard to find anywhere but within, burned-forest conditions. You would also find yourself hard pressed to see certain kinds of insects, toads, and plant species outside of a burned forest. In short, burned forests are magical places that seem to harbor plant and animal species and visual experiences found under no other forest condition. I ended up spending a good deal of my 38-year academic career trapsing around in, conducting research on, and then teaching about, burned forests. I'd like to share what I found and show you what birds are telling us about how current land management practices could be tweaked a bit to the benefit of all.

References

1. Oliver M (2006) Messenger. In: Thirst. Beacon Press, Boston
2. Hutto RL (1998) Using landbirds as an indicator species group. In: Marzluff JM, Sallabanks R (eds) Avian conservation: research and management. Island Press, Covelo, pp 75–92

2

Early Influences

I spent most days outside in my youth. I spent so much time exploring the wilds surrounding our neighborhood in La Crescenta, California, that my mom was compelled to tell me and my friends on more than one occasion to do something more useful with our time. "You boys do nothing but aimlessly wander," she'd say. But we were proud of our wanderings. To this day, I can still find where, in 1960 or thereabouts, I scratched "the explorers" into the newly poured cement curb at the top of New York Avenue just east of the debris-flow wash. We seemed to spend hours on most days hiking, catching western fence lizards, tarantulas, and the like, and building forts in huge laurel sumac bushes where we sometimes shot BBs at any poor bird that happened to alight nearby. With my buddy's Golden Guide at the ready, we would line up our quarry for post-mortem identification…Audubon's Warbler, Rufous-sided Towhee, Wrentit. It was awful in one sense, but educational in another. After all, real ornithologists relied on shotguns to

© The Author(s), under exclusive license to Springer Nature Switzerland AG 2025
R. L. Hutto, *A Beautifully Burned Forest*,
https://doi.org/10.1007/978-3-032-03180-8_2

identify birds in the field for most of the first half of the twentieth century. It wasn't until after the second World War that binoculars became more widely available to the public. That's not an excuse, of course. It's just, well, you had to be there as a 12-year-old.

Fires, or reminders of past fires, were also omnipresent in my neighborhood and became seared into the subconscious of all of us who lived there. Smokey Bear signs were visible reminders of fire danger, as were the miles of firebreaks etched into the San Gabriel Mountains, which towered over our patch of heaven (Fig. 2.1). John McPhee wrote elegantly in the New Yorker about the never-ending fires and subsequent floods in our neck of the woods [1]. Indeed, the periodic occurrence of fire was real. It seems like every 8mm movie my dad filmed of me and my sister in the neighborhood showed us pointing in sudden surprise at a fire in the chaparral-covered hills. Our friends accused us of repeatedly hamming it up for the camera, but we always offered the retort "no, really…there really *was* a fire!" On occasion we could jump on our bikes and get close to the real thing. I remember when embers from one chaparral fire blew into an adjacent neighborhood and destroyed a house at the top of Maryland Avenue. When we arrived on our bikes (it was undoubtedly easier to butt in among the fire trucks and fire hoses back then), the house was in full flame, and it was so hot at the edge of the driveway that I remember having to back off to escape the heat. So, like everyone else in America, I grew up fearing fire and the damage it wrought. Once I began to appreciate that structural fires are something entirely different from wildfires, however, my overall perception of fire began to change.

I began to take a more academic interest in fire after high school when I took classes at Glendale college, which was a mere 20-minute Honda Scrambler motorcycle ride from

Fig. 2.1   Even as recently as 1970 (when this picture was taken), an extensive network of firebreaks could be seen throughout the chaparral-covered San Gabriel Mountains, which towered just above the street where I grew up in La Crescenta, California

my house. Biology was one of those classes. I had avoided biology in high school for fear of having to pith frogs. Instead, I took chemistry and physics, which, along with Latin and French, were supposed to be the better choice for med school preparation anyway. My college biology teacher (a young gal with a master's degree from UC Santa Barbara) opened my eyes to two key ideas that came to influence my thinking about nature: the first was how natural selection leads to a directional change in the composition of individuals within a species when the species is subjected to a new environment. This insight emerged from an elegant experiment where bacterial colonies were blotted from one petri dish onto a series of new dishes. When an antibacterial substance was applied to the new dishes, there were surviving colonies in each dish. Amazingly, the surviving colonies occupied the same relative location in each dish, which meant that the resistant strains were not the result of randomly located colonies with mutations somehow spurred on or "forced" into existence by the altered environment. Successful variants were, instead, already existing at that one location in the original petri dish. It may not seem like much, but it revealed how and why natural selection acted as an important mechanism by which species change through time. Resistant bacterial strains, long antlers, cryptic colors, and all the other amazing things we see in nature come to be because early variants are favored and then further enhanced merely by selecting continuously and nonrandomly from among what already exists. No concept has been as useful or as meaningful to me throughout my life. Theodosius Dobzhansky captured the importance of natural selection perfectly when he said, "Nothing in biology makes sense except in the light of evolution" [2]. Honestly, natural selection is that powerful a concept.

The second thing that biology class did for me was to infect me with the idea that I might be able to major in something involving a good deal of time outside the classroom. This dawned on me after the whole class took a short walk into the chaparral one day. I found myself sitting right in a laurel sumac just like I was 12 again, but this time I learned not only the name of the plant, but also how it and the other chaparral plants had evolved leaf and root adaptations that allowed them to thrive in a fire-prone environment. It's amazing how the combination of immersion and a small dose of information can lead to such a large dose of appreciation for something. Anyway, after that point in college, I was all in with field biology!

The junior college experience was great. For $5 per semester, I took a slew of undergraduate courses that allowed me to earn an Associate of Arts degree before choosing to continue my undergraduate career at UCLA, where I then had to pay a whopping $107.50 per quarter to rub shoulders with some of the best field biologists in the country. I was soon visiting different-aged patches of chaparral to learn about the process of plant succession following several previous fires in Topanga Canyon as part of a class in ecology. I also found myself trapping desert rodents in a mammalogy class and then getting to design and conduct my own independent research project on woodrats in the Granite Mountains of the Mojave Desert as part of a field ecology class. From there it was off to Northern Arizona University for my M.S. and then back again to UCLA for my PhD. Both of my graduate projects involved an attempt to uncover how similar species differ to promote coexistence in the same place—desert rodents for my masters and wood warblers for my dissertation. Importantly, both projects involved extensive field work. I established study sites across the desert southwest for my master's project. For my dissertation work after that, I managed, during each of several years, to visit survey locations that were scattered across every state within western Mexico in winter, the Chiricahua Mountains in Arizona in spring, the Sierras of California and the Tetons of Wyoming in summer, and the Chiricahuas again in fall. Extensive travel most certainly helped Darwin see patterns that the stay-at-homes missed, and my own research travels undoubtedly helped me detect patterns of association between critters and specific environments as well.

Travel between the Neotropics and the Nearctic was one thing for a graduate student who rarely set foot on campus but was going to be quite another thing for an academic saddled with teaching, research, and service. How was I going to be able to maintain an approach to research that involved extensive travel? Two things made my research style possible: a university operating under the quarter system and a supportive set of colleagues who understood that there are sometimes unconventional travel needs associated with field work.

References

1. McPhee J (1988) Los Angeles against the mountains—I. The New Yorker
2. Dobzhansky T (1973) Nothing in biology makes sense except in the light of evolution. Am Biol Teacher 35:125–129

3

All In

I was tickled to discover that the University of Montana was on the quarter system when I arrived there. The gift of three school-year teaching blocks allowed me to teach in the fall and spring quarters and to head back to Mexico with students during the winter block to continue studying how migratory birds fit into the subtropics during the nonbreeding season. Nevertheless, heading off-campus to do research for a couple of months each winter had its challenges, and when I learned that the university planned to switch to a semester system, my winter research program was only going to get harder to maintain. I needed to shift my main research focus sooner or later and, wouldn't you know, I got more than I wished when the fires of 1988 burned more than a third of Yellowstone National Park. Ever since the Pattee Canyon fire, I had really wanted to look deeper into the relationship between fire and birds I witnessed there; this was a chance to conduct some formal fire research.

R. L. Hutto, *A Beautifully Burned Forest,*
https://doi.org/10.1007/978-3-032-03180-8_3

While the historic Yellowstone fires were burning, NBC News anchor Tom Brokaw told the American public that our first national park was (and I quote) "under siege," as if the natural world was at risk of being wiped out forever. Public perception of forest fires has not changed appreciably since then—I can read opinion pieces in our local newspaper on most any day of the week and see the same sentiment expressed. For example, Tom Partin, Montana representative for the American Forest Resource Council, wrote in an opinion piece for the Missoulian (17 September 2021) that "Hundreds of fires have burned across the West this summer, charring millions of acres and causing death, destruction and leaving our national forest landscapes in waste." You get the idea.

My limited field experience in burned forests suggested something quite different from what reporters were saying in 1988 (and what they are still saying today). I needed to investigate in a more formal way whether those recently burned areas had become seas of death and destruction, as described on the evening news, or whether there was, instead, a special kind of natural transformation that occurred in the park and elsewhere.

It was primarily wanderlust that pushed me early in my research career to visit field sites across every state in western Mexico, and that certainly contributed to my preference for using a comparative approach in my scientific explorations from then on. While I was in Mexico, my desire to compare what I found from one place to the next was so strong that I even continued south from Colima after being warned by a Mexican mechanic that I should head straight back to the US and not head any farther south. I had just completed a series of surveys up and down the nearby volcano when my VW squareback engine began to tick loudly. After removing the valve covers to investigate, I found that one of the rocker arm bolts had loosened. Unfortunately, it wouldn't tighten down; it was stripped. Luckily, I had a tow bar and found a taxi (probably the only one in Mexico) with a trailer hitch. After we made it to an auto shop, I was elated when that mechanic first said, "No problem; we can just ratchet in an oversized bolt!" He added the warning I alluded to above, "This is a temporary fix and should get you back to the US and no more." I then ignored his advice and headed south for Mexico City and beyond. I'll be darned if the bolt didn't start loosening again after I had driven about as many miles as I would have if had I headed north to the border. Luckily, it held when I tightened it again, so I was able to continue exploring the wilds of western Mexico all the way south to Oaxaca.

The wanderlust associated with my Mexican research allowed me to see that exceptionally large mixed-species flocks occurred throughout western Mexico, but only in the highlands. Flocking behavior was widespread and an integral part of the way migrants and residents coexisted in those forests in winter. A typical flock might consist of more than 30 species and 100 individual birds all moving together through the woods. If I had observed such a large flock in only one location, I might have passed it off as an amazing experience, but also a fluke and no more. Wanderlust, however, allowed me to observe the same thing over and over in different places, and the geographical extent of the flocking phenomenon made its significance sink in. I subsequently described these flocks in the literature as being common in the highlands and larger than mixed-species flocks occurring anywhere else in the world [1]. I didn't make this claim after having surveyed birds everywhere else in the world, of course. I used the next best substitute—published reports of flock compositions and sizes worldwide. The comparison allowed me to conclude that western Mexico was in a class by itself.

The same degree of wanderlust (and interest in chasing after birds) positioned me perfectly after the Yellowstone fires of 1988, which garnered plenty of national attention. Those fires were the result of a combination of unusually high winds and drought conditions (the driest summer in the park's history), which led several naturally occurring fires to merge into what ended up burning a combined 1.5 million acres (0.6 million ha) across the Greater Yellowstone Ecosystem and maybe another half million ha across 40 or 50 additional fires scattered throughout western Montana (Fig. 3.1). The explorer in me knew that I had been handed the opportunity of a lifetime—I could visit them all!

By conducting bird surveys in dozens of independent fires, I could achieve an unheard-of level of natural experimental replication and could get a good feel for what, on average, happens following a fire event. I needed to score financial support in rapid fashion and, fortunately for me, the National Geographic Society gave me more than enough to get underway. I assembled an experienced field crew, grabbed my binoculars, and found myself trapsing across Montana during the next few field seasons.

Every day in the field seemed to bring an entirely new experience. I remember how wonderful it was to feel like we had the whole thing to ourselves. No other scientists were looking at the things we were looking at—none. Even better, we could hike all day in any one of those fires and never see a soul. It was heaven. As the days passed, patterns began to emerge. These patterns of nonrandom habitat use were so strong that even without conducting a formal statistical analysis, I knew we were witnessing something special. Perhaps the most obvious thing that would have hit anyone with geographically diverse birdwatching experience is that

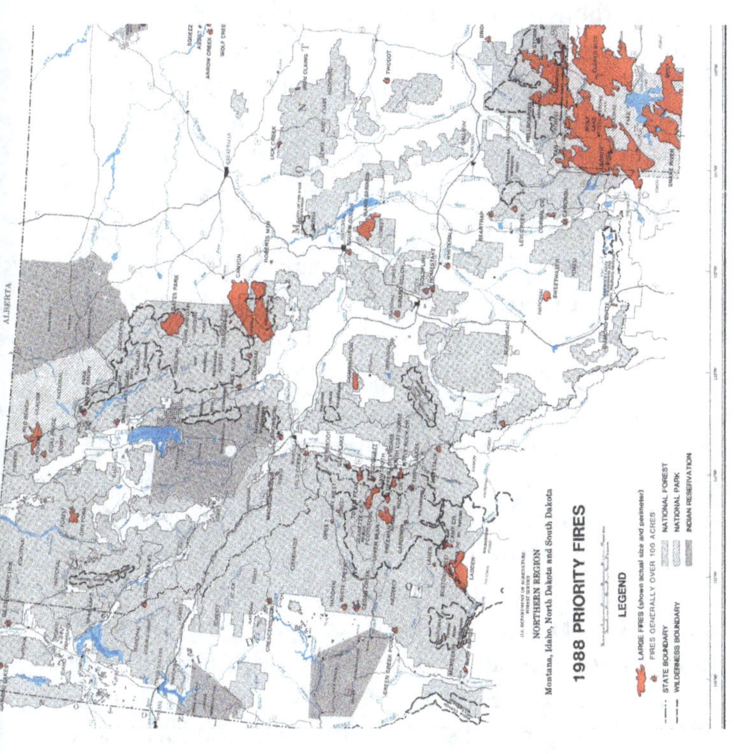

Fig. 3.1 This USFS Northern Region map of the 1988 fires (highlighted in red) reveals the gold mine of potential independent study areas that landed in my lap that year. I conducted bird surveys in most of the burned areas pictured here in the ensuing years

in a burned forest, one can see species in abundance that are nowhere near as abundant outside a burned forest. The burned-forest condition made it easy to see Mountain Bluebirds, Clark's Nutcrackers, and woodpeckers of every stripe, just as I witnessed after the Pattee Canyon fire a decade earlier. There seemed to be a special kind of magic associated with burned forests.

Even though my experience conducting formal bird counts across the West told me right away that these burned forests held combinations of bird species not found elsewhere, I needed quantitative data to test my qualitative impression. Today, one could go to eBird, a citizen-science generated database, to get rough estimates of the probabilities of occurrence for any bird species across a wide range of unburned vegetation types to compare with the burned forest data, but there was no eBird back then. I couldn't submit a manuscript to a peer-reviewed journal and simply say "Believe me, these places are cool!" So, I had to find a way to demonstrate that what we found in burned forests can be found in no other vegetation type or condition in the region. I used the same trick I used to argue that mixed-species flocks are larger and more diverse in western Mexico than anywhere else in the world—I used published data from outside burned forests as a basis of comparison with my own data. Specifically, to compare bird community composition typical of burned forests with bird communities in unburned vegetation types, I amassed data from numerous breeding bird surveys that others conducted across a variety of Rocky Mountain vegetation types and were already published in either American Birds or Audubon Field Notes. I then used results from that literature survey to tally, for each species, the proportion of studies in which it was present (its probability of occurrence) across each of 12 different vegetation types, with one of the "types" being burned conifer forest. It was somewhat of an apples and oranges comparison because the methods used to conduct breeding bird surveys were not the same among studies and were not based on the same point-count survey protocol I used in my own work in the burned forests.

Nonetheless, I calculated the probability of occurrence of different bird species within each of the 12 vegetation types. An eye-opening thing emerged from this analysis. No fewer than a dozen species appeared to be more abundant in burned forests than in any of those other unburned major vegetation types! Most importantly, twelve species were nowhere more abundant than they were in burned forests, and several of those species (most notably, the Black-backed Woodpecker) were not only more abundant in than anywhere else but they seemed to be relatively *restricted* in their distributions to burned forest (Fig. 3.2). Let that sink in. How could a bunch of bird species be so abundant in, and even relatively restricted to, the very kind of standing-dead, burned forest that the news media repeatedly tells us is the unnatural product of a forest out of whack?

To this day, the iconic Black-backed Woodpecker is most extreme example of fire-dependency in any bird species, worldwide. In the 1995 paper where I first described this result, I wrote, in as strong a language as could be expressed in a scientific publication at that time, "I believe it would be difficult to find a forest-bird species more restricted to a single vegetation cover type in the northern Rockies than the Black-backed Woodpecker is to early postfire conditions" [2]. The unique association of this woodpecker with burned forests was even noted (but never fully appreciated) long ago in Alexander Bent's (1939) classic Life History of North American Woodpeckers where he highlighted the habitat relationship by quoting Manly Hardy, who wrote that while visiting a fire-killed forest, he shot the heads off

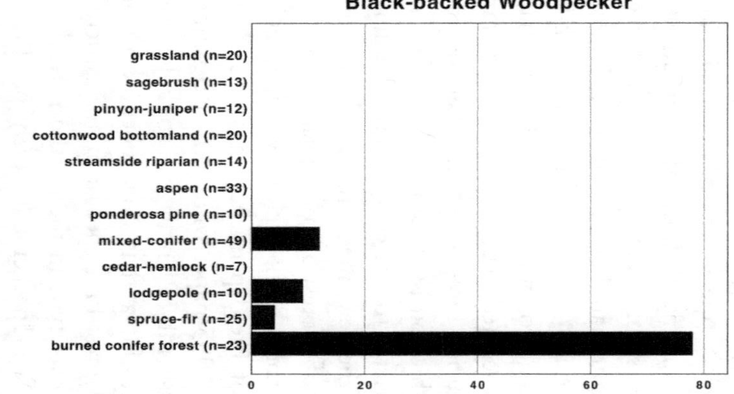

Fig. 3.2 For each of twelve Rocky Mountain vegetation types, bars show the percentage of published breeding bird surveys that included the Black-backed Woodpecker. This graph is based on the original data I used in a 1995 publication to show that the woodpecker species occurs nearly exclusively in burned forests within the region

six Black-backed Woodpeckers within a few minutes when he was short of material for stew! I can't overstate the importance of the Black-backed Woodpecker's pattern of habitat use. The only way a species could be restricted to burned forests is if it evolved in the presence of that forest condition for a very long time and concurrently evolved a behavior that limited its habitat breadth to that specific forest condition.

How is it even possible for a plant or animal species to become relatively restricted to a forest condition that is ephemeral in nature? And how in the heck could a Black-backed Woodpecker even find a newly burned forest? I'll get to the first question later, but concerning the second question, if you go outside in the mountainous West in mid- to late-summer in almost any year, you can stand in one place, spin around, and see smoke rising from a fire somewhere nearby. Therefore, it's probably not too hard for a bird to evolve the ability to discover newly burned forest as part of its dispersal behavior. We now know from radio telemetry studies that young birds are more likely to disperse than adults, and that young can move 30 miles to a new burn [3]. The wonders of birds in burns never cease to amaze me.

I was on an emotional high being surrounded by so many woodpeckers, bluebirds, buntings, tanagers, and the like on nearly a daily basis during the first few summers following the 1988 fires. My future wife, Sue, and I spent many of those days together surveying birds. She would head out in one direction and I'd head in the opposite, and by lunchtime we couldn't wait to reconnect to compare the morning's notes. The routine was a familiar one to us, having spent months doing the same thing across western Mexico while studying the distributional ecology of migratory landbirds in winter. On the evening before we were to survey birds in the 1988 Corral Creek fire south of Ennis, Montana, we set up camp on a nice bench overlooking the Madison River Valley below. I was enjoying Sue's companionship so much that I proposed marriage then and there. Looking back, I could have picked a better place and time, but she agreed to go all in anyway. She tells others that I proposed in a cow pasture. I admit, cow patties were scattered here and there, but I swear we were camping on the open range and not in a fenced pasture!

References

1. Hutto RL (1987) A description of mixed-species insectivorous bird flocks in western Mexico. Condor 89:282–292
2. Hutto RL (1995) Composition of bird communities following stand-replacement fires in northern Rocky Mountain (U.S.A.) conifer forests. Conserv Biol 9:1041–1058
3. Siegel RB, Tingley MW, Wilkerson RL, Howell CA, Johnson M, Pyle P (2016) Age structure of Black-backed Woodpecker populations in burned forests. Auk 133:69–78

4

Research by Exploration

Many who read the first paper I wrote about birds in burned forests were skeptical of my claim that Black-backed Woodpeckers were relatively restricted to burned forests throughout the northern Rocky Mountains. I even sensed doubts in some corners about whether the data I amassed were adequate to show that Black-backed Woodpeckers were rare everywhere outside burned forests. After all, the pattern was based on data collected from only 12 different vegetation types and involved widely different field methods. I was confident from the breadth of my field experience that the patterns I described were not an artifact of combining results from different survey methods. Still, I needed to find a way to compare survey data collected using the same survey methods across a more complete range of vegetation types.

Fortunately for me, a nationwide initiative to create a collaborative Neotropical Migratory Bird Conservation Program had just been hatched at a large multi-agency and

R. L. Hutto, *A Beautifully Burned Forest*,
https://doi.org/10.1007/978-3-032-03180-8_4

NGO workshop in Atlanta in December of 1990. The workshop was convened to address declines in migratory landbirds. We were told that it was the first time the heads of four federal agencies (U.S. Fish and Wildlife Service, U.S. Forest Service, Bureau of Land Management, and National Park Service) had gathered in one room; it was a significant meeting. Even more fortunately for me, Alan Christensen of the USFS Northern Region office in Missoula knew I had worked throughout the West on migratory landbirds. He reached out earlier that same year to ask if I'd be interested in discussing how the regional office might best spend an earmarked budget allocation that they anticipated receiving because of the partner agreement the USFS was expected to sign following the 1990 Atlanta meeting. That was a no brainer. I collaborated with Sallie Hejl, another avian ecologist who worked just off campus at the USFS Rocky Mountain Research Station, to design what turned out to be just what I needed—a region-wide bird monitoring program where we would conduct formal bird surveys using the same point-count method across nearly every vegetation type and vegetation condition in existence in the northern Rocky Mountains.

The most popular method used for bird surveys at the time was a line-transect method that John Emlen had refined to a point where his name became fully attached [1]. He introduced refinements to address a major concern that the proportion of individuals detected within a fixed width on either side of a transect line will differ from one species to the next. Some species can be detected over long distances and others only when they are close by. For example, an observer might detect only 50% of the Brown Creepers within 100 meters of a transect line but 100% of the American Robins. The density of creepers would, therefore, be severely underestimated relative to robins unless one limited bird detections to those within a relatively small transect width. To avoid having to use a single, narrow width (which would require throwing away the more distant bird detections), Emlen proposed using all detections to determine, for each species, the species-specific distance from the transect line that ought to be used as a cutoff when calculating its density. The method may have allowed researchers to better compare bird densities across species, but the method did nothing to solve a more fundamental problem with transect surveys—any transect will pass through multiple vegetation types, which makes it hard to determine whether a species is associated with (or restricted to) any one of those vegetation types. For those, like Sallie and I, who wanted to understand habitat relationships, transect surveys needed to be replaced with something else. I had been struggling with this issue while conducting bird surveys in Mexico when my graduate students (Paul Hendricks and Sandy Pletschet) and I decided the best approach was to abandon the transect as a sample unit and, instead, stop for a period (10 min) at largely independent points along the transect and then use each point as a sample unit for data analysis. In contrast with a cumulative tally of data from a mile-long transect, a single point (and all the birds detected from that point) could be more easily linked with a single vegetation type or land condition. And we could obtain 10 times the number of independent samples in the time it would take to complete a single transect-based survey. We ultimately published a description of the fixed-radius point-count method we homed in on [2]. Variations on the point-count theme were soon gaining enough traction that a large group of ornithologists agreed to meet in 1991 to establish a standard for conducting this relatively new point-count method in bird survey work [3]. That standard is still in use today.

Sallie and I implemented the basic point-count survey method for our bird monitoring program. In addition to documenting vegetation information around each survey point, we devoted considerable effort (before the availability of GPS units) toward mapping (on topographic quadrangle maps) survey point locations so we could gather landscape-level context data from aerial photos as well. Our effort became known as the USFS Northern Region Landbird Monitoring Program. It allowed us to hire around 20 field technicians per year to conduct bird surveys throughout northern Idaho and Montana, and to collect what was also precisely the kind of data I needed to compare results among vegetation types. A few bird observatories had started long-term landbird monitoring programs in some western states at about the same time, but we were the only university-based operation and the only monitoring program focusing on habitat relationships rather than population trends. Our effort produced what was probably the first long-term, region-wide, habitat-based, georeferenced landbird monitoring program in the nation. I was able to hire a small staff (Jock Young and John Hoffland and, soon after, Amy Cilimburg, Kristina Smucker, and Anna Noson) to coordinate tasks associated with finding, hiring, training, and supervising field crews, and with handling data management and data analysis. I can't overstate how lucky I was to have those folks in the fold early on. Over the ensuing 20 years we amassed enough data to look at the distributions of all landbird species across nearly three dozen vegetation types using the same point-count survey method everywhere.

I was eventually able to combine results derived from our region-wide point-based surveys with my earlier burned-forest survey data. The subsequent comparison of bird abundances across vegetation types confirmed just how

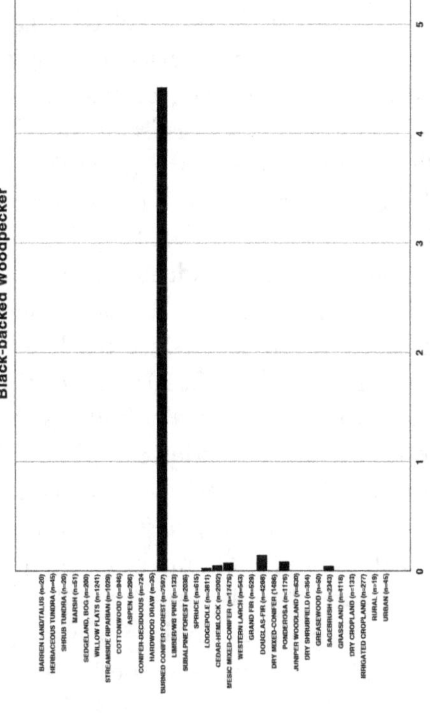

Fig. 4.1 The percent occurrence of Black-backed Woodpecker from point counts conducted within 32 vegetation types in western Montana between 1988 and 2008. More than 100 counts were conducted in each vegetation type and a total of more than 20,000 points were visited. These striking results underscored my earlier suggestion that the Black-backed Woodpecker is nearly restricted in its distribution to burned conifer forests

restricted the Black-backed Woodpecker was to burned forests [4]. Indeed, the abundance of the Black-backed Woodpecker in burned forests in comparison with its abundance across a wide range of Rocky Mountain vegetation types stuck out even more dramatically than when I used data from a literature review of relatively few vegetation types (Fig. 4.1).

When research results boil down to what can be presented in a simple bar chart, the effort involved is largely hidden. The representation of a bird's relative abundance across many vegetation types hides the fact that an across-year total of nearly 200 top-notch field assistants were hired and trained to collect the data needed to inform the story. We hired some of the best bird watchers in the nation to help in the years it took to amass the data (Fig. 4.2). I was also lucky to have more than a few graduate students work on their own questions surrounding the ecology of burned forests during that time. In the years since the Landbird Monitoring Program ended in about 2010, additional data have only served to corroborate our earlier findings [5], as have research projects and surveys conducted by Point Blue and the Institute for Bird Populations across the Sierras of California. Some of the most recent published work from the Sierras involved the use of radio-marked birds that I mentioned earlier, revealing that dispersing juvenile Black-backed Woodpeckers immigrate selectively into burned forests above all other vegetation types and conditions there too [6].

The association between this woodpecker species and burned forests is not absolute, of course. No species on planet Earth can be said to occupy one and only one habitat type or condition. That said, throughout the West, the Black-backed Woodpecker is still relatively restricted to a single forest condition—severely burned forest. They occur

Fig. 4.2   Field crew getting ready to conduct bird surveys in a burned forest within Glacier National Park in 2006

in unburned, green-tree forests as well, but at much lower densities. I used to tell people "I will give you $10,000 if you can show me a Black-backed Woodpecker in a western green-tree forest, provided you give me $10,000 for every Black-backed Woodpecker I show you in a severely burned forest; if you take that bet, I will walk away a rich man!" We've always known that small numbers of black backs occur in green forests, especially within a mile or two of a recently burned forest. That's not news and it's largely irrelevant. A relatively tiny abundance of black backs in green forests in no way negates the significance of its relatively high abundance in burned forests.

References

1. Emlen JT (1971) Population densities of birds derived from transect counts. Auk 88:332–342
2. Hutto RL, Pletschet SM, Hendricks P (1986) A fixed-radius point count method for nonbreeding and breeding season use. Auk 103:593–602
3. Ralph CJ, Sauer JR, Droege S (1995) Monitoring bird populations by point counts. USDA Forest Service General Technical Report PSW-GTR-149:1–181
4. Hutto RL (2008) The ecological importance of severe wildfires: some like it hot. Ecol Appl 18:1827–1834
5. Hutto RL, Hutto RR, Hutto PL (2020) Patterns of bird species occurrence in relation to anthropogenic and wildfire disturbance: management implications. For Ecol Manage 461:117942
6. Stillman AN, Lorenz TJ, Siegel RB, Wilkerson RL, Johnson M, Tingley MW (2021) Conditional natal dispersal provides a mechanism for populations tracking resource pulses after fire. Behav Ecol 33:27–36

5

Interlude I

Before the fires of 1988, I took walks regularly in the burned forest in Pattee Canyon because it was only a few miles from where I lived in town. It was a place where the sounds of trains and cars and garbage trucks and dogs were silenced by some distance, and where I could really listen to sounds of the forest. By the first summer following the fire, things really began to take a hold on me. Emanating from this burned forest were subtle creaks and cracking sounds in the bark layers as the trees warmed up, the buzz of native bumblebee pollinators, and by mid-morning, chewing sounds coming from wood-boring beetle larvae feeding on the wood beneath the bark of the thick-barked trees. I found that if I approached the sound-emitting location, the chewing sound stopped, which required that I freeze in position and wait patiently for the sound to resume once again. Once I got close to the part of the tree that seemed to be harboring the sound, I could flick away the bark with a knife and find the grub doing the chewing (Fig. 5.1). I

R. L. Hutto, *A Beautifully Burned Forest*,
https://doi.org/10.1007/978-3-032-03180-8_5

Fig. 5.1 Flat-headed woodborer larva captured from beneath the bark of a severely burned and blackened Douglas-fir tree

figured if it's that easy for a human to find a beetle larva, then it must be a piece of cake for a woodpecker that has evolved the ability to really listen in. In some spots I could hear several woodpeckers digging for beetle larvae at the same time, making the place sound like a bunch of carpenters working in the woods. Some woodpecker species, like the Pileated Woodpecker, did a lot of flaking, while others, like the Hairy Woodpecker, spent more time digging. Then there were the woodpecker calls, which carried across the entire canyon. The "peek" call of a Black-backed Woodpecker, which you can replicate by whistling through your teeth while saying peek, is as unique as the "peet" call of a Downy Woodpecker or the "pweet" of a Hairy Woodpecker, and subtly different from the "pqueet" of an American Three-toed Woodpecker. After the burned forest had been visited by woodpeckers for a year or two, the Douglas-fir trees began to assume one of my favorite visual outcomes—woodpecker digging patterns etched on blackened trees (Fig. 5.2).

Fig. 5.2 A beautiful sign that woodpeckers had previously dug extensively for beetle larvae on a blackened Douglas-fir tree

Perhaps the most amazing thing I eventually discovered there was the way different species of woodpeckers could be found foraging in mixed-species flocks in winter. This finding is remarkable to me and has still not been described in the primary literature. I remember raising my binoculars at one point and being blown away by how I could see 6 species of woodpeckers in the same field of view! They (Pileated Woodpecker, Northern Flicker, Downy Woodpecker, Hairy Woodpecker, American Three-toed Woodpecker, and Black-backed Woodpecker) could be seen simultaneously pounding away on nearby trees. Wow…I happened upon yet another unusual mixed-species flock phenomenon. It's probably true that few people had ever birdwatched in burned forests back then, and even fewer had done so in the middle of winter. My discovery was possible only because the fire had burned so close to town, making my visits in the middle of winter relatively easy. These woodpeckers foraged together as a loose group, which enabled them to gain the benefit of protection from avian predators through a built-in early warning system—it's hard to escape detection from so many eyes on the lookout. As they extracted wood-boring beetle larvae that had exploded in numbers following the fire, it became apparent that the tree species and heights at which the different woodpecker species fed were noticeably different as well.

As if that weren't amazing enough, the behavior of Calliope Hummingbirds several years later became yet another astounding thing I discovered in that burned-forest environment. Scattered willows had grown to be 5 m tall, and dead twigs atop those willows served as perfect perch sites for male Calliope Hummingbirds. I started collecting data on the time budgets of those hummingbirds because they were easy to spot, and because as many as 8 hummers could be watched from a single location. Several students and I spent mornings there with stopwatches to document that any given male spent about 80% of his time perched on as few as 3 perch sites. The rest of the time the male fed on honeysuckle and other nectar-producing plants and periodically performed dramatic dive displays where he flew upward to maybe 30 m above the ground and then dove straight down to produce a "brzzzt" sound from his spread tail feathers at the bottom of the dive after which he flew straight up again to perform another dive or two. Female hummingbirds (and not uncommonly Chipping Sparrows, Dusky Flycatchers, and Warbling Vireos) were usually located just below the bottom of the dive and were clearly the objects of these dive displays. The aerial displays had been well described, but when I first started watching these birds in 1983, nowhere in the literature had anyone reported so many birds performing so close to one another. Neither had anyone described what happens when the male slowly descends after several dives and performs what has since come to be known as a "shuttle display" in front of the female's face. The first time I saw this I was awestruck. In a shuttle display, a male hovers right in front of the female and spreads his chin feathers (his gorget) so that his bill appears to be a sword directed forward from within a dish-like amphitheater of feathers (Fig. 5.3). He then shuttles his hovering body back and forth, keeping the tip of his bill in place while producing a wing-induced noise sounding exactly like a bumblebee buzzing in a flower to shake pollen loose. This display is followed by a quick copulation with the perched female before she then flies off. This shuttle display (what I originally labeled a "bumblebee buzz" when I wrote

Fig. 5.3 The Calliope Hummingbird performs amazing courtship displays in severely burned conifer forests amid willow shrubs that develop in the understory a decade or two following fires. (Photo Gordon Brown)

about this [1]) had not yet been described either. I came to conclude that the early successional, open, shrubby environment that develops following severe fire (or, in some cases, heavy logging) is sometimes perfectly suited to accommodate potential display sites for many males at the same time. The simultaneous aerial displays of so many males may function much like a "lek" or display site does for, say, Sage Grouse, where many males display at the same time, but only one or two mate with the visiting females. The females go on to build nests and raise young on their own, so they have little need for males other than to fertilize their eggs. That allows for the evolution of this unusual behavior in special locations where the males can aggregate and compete for the chance to mate. Recently burned forests are truly magical places.

Reference

1. Hutto RL (2014) Time budgets of male Calliope Hummingbirds on a dispersed lek. Wilson J Ornithol 126:121–128

6

From Fire Events to Fire Regimes

I've been writing as if all forest fires create postfire conditions just like the ones I've been describing, but that's not at all true. There is a good deal of complexity associated with fire, and the biological effects of fire depend on *how* a fire burns. One kind of fire can benefit, while a different kind of fire can hurt, any single species. Therefore, results from different studies of fire effects on a plant or animal species can produce different results simply because the nature of the fire differed across studies. In fact, in an early review of published studies of the effects of fire on birds, we found that the studies often yielded contradictory results [1]. Indeed, we even classified several species as "mixed responders," implying that they might respond one way to a fire this time and another way the next. In hindsight, it's clear there is no such thing as a "mixed responder;" birds respond quite predictably to fire. Results that differed among studies were merely reflecting differences in the kinds of fires studied— the *kind* of fire matters to organisms.

R. L. Hutto, *A Beautifully Burned Forest*,
https://doi.org/10.1007/978-3-032-03180-8_6

In what ways do fires differ from one another? There are several key physical attributes of a fire disturbance event, which might affect the organisms that live there. Any individual fire event (indeed, any disturbance event) can be characterized by its (a) magnitude, (b) patchiness, (c) extent, and (d) timing. The magnitude of a disturbance is a measure of the force of the disturbance (intensity) measured as energy released per unit area per unit time. It also refers to the physical damage caused by the disturbance (severity) measured by the effect on organism, community, or ecosystem properties. Note from these definitions that fire intensity is not the same thing as fire severity, but they are well correlated. The USDA Forest Service uses a satellite-based, burned area reflectance classification scheme to label areas within a fire perimeter as belonging to unburned, low-, medium-, and high-severity classes for national-level support of Burned Area Emergency Rehabilitation (BAER) activities. These classes are generally the result of ground, surface or understory, and crown or stand-replacement fires which, in turn, reflect <25%, 25–70%, and >70% canopy tree death. Patchiness refers to how uniformly severe the fire was; extent is simply the area burned within the fire perimeter; and timing can refer to time of year, time of day, or maybe even time since the last fire burned in a specific place. The effect of a disturbance event on biological systems will depend, in part, on the values of these single-event variables, so it is important that studies of disturbance effects include information about them. In terms of fire effects on living organisms, perhaps the most important is fire severity. It might surprise you that, even as recently as 20 years ago, very few scientists who studied the ecological effects of fire included information about the nature of the fire that produced the conditions being studied. That's why some of these earlier studies of fire effects yielded results that seemed to contradict one another.

That's also why I need to be careful to define the kind of fire I'm writing about in this book. I am writing primarily about fire in mixed-conifer forests of the West, with a focus on the Northern Rockies, where I live. There, fire is not only a naturally occurring, periodic agent of disturbance, but fires are generally *severe* disturbance events (as they are in most mixed-conifer forests elsewhere in the West). The amazing postfire response of plants and animals I have already referred to is restricted to these kinds of tree-killing, severe forest fires.

Besides severity and the other single-fire attributes of magnitude, patchiness, size, and timing, we must also appreciate that the historical pattern of repeated burning (fire frequency) also varies from place to place and forest type to forest type. Different places experience not only different kinds of fire, but also different fire frequencies when viewed across long periods of time. Fires that burn at different frequencies will provide very different environmental backdrops for plant and animal evolution. Suppose fires historically tended to burn infrequently (say, every 30 years, on average) in a particular geographic location and vegetation type. Over many hundreds of years, a plant growing in that environment might evolve a developmental program that instructs it to grow for a fairly long period (say, 10 years) before seeding profusely. Now imagine that the same area begins to experience fires in unnaturally rapid succession— one after 5 years, another 2 years after that, and yet another 3 years after that. You can see the problem a plant faces in the hypothetical scenario of needing at least 10 years to set seed—fire can be too frequent for a plant's own good if it can't reproduce successfully. Unfortunately, the hypothetical problem of too-frequent fire is all too real in California's chaparral shrublands, where the invasion of non-native grasses and the decline in reproductive success of native plants is a serious concern [2]. This problem illustrates that the effects of a fire depend on not only the fire's magnitude, patchiness, extent, and timing, but also with how long it has been since the last fire occurred in that location. For this reason, fire scientists use fire frequency as another important fire descriptor. By looking at fire frequency along with averages of each of the four variables used to characterize any single fire event, scientists can build a picture of how often multiple fire events tend to occur in the same place across time. The combination of information on the frequency of single-fire events along with the average magnitude, patchiness, extent, and timing represents a *fire regime*—a kind of statistical summary of how multiple fires burn in a single location or vegetation type over time. A fire regime constitutes an important environmental backdrop against which plants and animals evolve over long periods of time. In any given place, the historically important kind of fire regime is what land managers ought to be striving to maintain.

The magnitudes of some regime descriptors are well correlated, which means if you know the value of one descriptor, you automatically know the value of another as well. For example, the strong inverse relationship between fire frequency and fire magnitude means a place that experiences frequent fire will, on average, also experience low-severity fires. The values of these descriptors are also reasonably predictable for any given vegetation type and location. Consequently, various authors have proposed systems for using the descriptors to label different fire regimes. Similar forest types tend to experience similar fire regimes, and most schemes boil down to three basic categories: understory (frequent, low-severity), stand-replacement (infrequent, high-severity), and mixed-severity (mixed frequency and severity).

Mixed-conifer forests experience repeated disturbance that always includes at least some severe, crown fire. Crown fires act as *overstory* disturbance agents where large swaths of overstory vegetation are burned and many trees are killed. Indeed, most western conifer forests routinely experience crown fire events, although the extent and frequency of such fires varies with forest type and geographic location. Open-grown pine forests in the Southwest, for example, experience much less severe fire, on average, than do most other western conifer forests. Frequent, low-severity, understory fires (repeated *understory* disturbance events) keep ponderosa pine forests suspended in a mature-looking, green forest condition. Ponderosa pine forests are frequently said to persist in a "steady state" because they appear relatively unchanged through time. It's also important to note that understory fires do not promote a marked turnover in the kinds of plants and animals that occupy a site following fire because such fires do not yield widespread overstory plant death (the key ingredient of a severe disturbance event, as I'll describe shortly). Therefore, understory fire events in open-grown ponderosa pine forests are qualitatively different from the overstory fire disturbance events (Fig. 6.1). Overstory fires burn more extensively and more frequently in mid- to high-elevation mixed-conifer forests north of the southern Rockies and north of the southern Sierras than they do elsewhere. Those forest types constitute most of the conifer forest cover throughout the West and they are the forests that have precisely what it takes to create those unique on-the-ground conditions that support the most fire-dependent species I have been writing about.

The categorization and simplification of fire regimes into just three types is useful, but it hides much of the variation

Fig. 6.1 An overstory crown fire (top series) creates a magically transformed, blackened forest of standing dead trees (left to right), while a low-severity, understory fire creates a forest that is pretty much the same as it was before the fire. (Figure modified from Davis et al. [4])

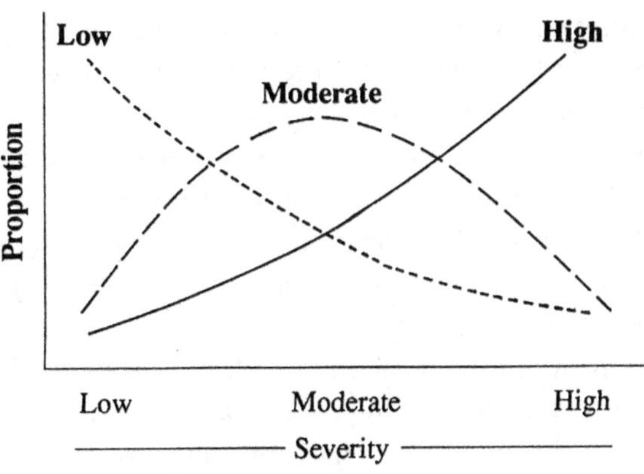

Type of Fire Regime

Fig. 6.2 This graphical illustration of three fire severity regime types captures the idea that every fire regime has measurable proportions of all fire severity classes. (Figure from Fire Ecology of Pacific Northwest Forests by James K Agee. Copyright 1993 by the author. Reproduced by permission of Island Press, Washington, DC.)

in fire behavior and fire outcomes among vegetation types and across spatial scales. In his classic 1993 text entitled Fire Ecology of Pacific Northwest Forests, Agee included a figure (reproduced here as Fig. 6.2) that captures the most important lesson one can learn about fire regimes—most all fire regimes consist of all fire severities. Examine Agee's simplified graph depicting just three fire regime classes along the x-axis. Specifically, draw a line up from within any fire regime class to see that all three fire severities are associated with any regime, but in different proportions from other fire regimes. Thus, all fires have each of the three severities, even though the *average* varies from mild to severe across the x-axis. Thus, saying that a certain vegetation type experiences a fire regime labeled by one kind of fire that occurs there (e.g., low-severity fire regime) hides the fact that severe fires occur, albeit rarely, even in low-severity regimes. They also hide the fact that different severities occur at different spots *within* every fire.

Relatively few kinds of forests experience fire regimes that fall near the extremes along the fire-regime axis depicted on the graph (either exclusively low-severity, understory fires or exclusively stand-replacement, overstory fires). Instead, roughly 85% of all forested lands within the western US fit well within the intermediate mixed-severity category, which includes low-, moderate-, and high-severity fire in surprisingly equal proportions across vegetation types and biophysical settings [3]. Sometimes I wonder if a mixed-severity fire regime might even characterize ponderosa pine forests better than the frequently used "understory" or "low-severity" fire regime labels. Giving a forest the "low-severity" fire regime label leads one to believe that the forest type experiences frequent, low-severity fires and nothing else, but that concept is wrong. At least in the northern Rockies, where frequent low-severity surface fires were historically common in ponderosa pine forests, ecologists have shown that infrequent high-severity fires also occurred historically. Think about it…if you accepted mixed-severity (rather than frequent, low-severity) as the nominal fire regime for ponderosa pine forests, then you'd probably view some of the recent severe fire events as natural but rare occurrences!

References

1. Kotliar NB, Hejl SJ, Hutto RL, Saab VA, Melcher CP, McFadzen ME (2002) Effects of fire and post-fire salvage logging on avian communities in conifer-dominated forests of the western United States. Stud Avian Biol 25:49–64

2. Park IW, Jenerette GD (2019) Causes and feedbacks to widespread grass invasion into chaparral shrub dominated landscapes. Landsc Ecol 34:459–471
3. Hutto RL (2021) Fire and climate change: a comment. Front Ecol Environ 19:86–87
4. Davis KT, Peeler J, Fargione J, Haugo RD, Metlen KL, Robles MD, Woolley T (2024) Tamm review: A meta-analysis of thinning, prescribed fire, and wildfire effects on subsequent wildfire severity in conifer dominated forests of the Western US. Forest Ecology and Management 561:121885

7

Fire as an Agent of Disturbance

Western conifer forests undergo dramatic seasonal changes, from the dead-of-winter, seemingly lifeless quietude (save a chickadee-nuthatch flock here and there), to the springtime cacophony of bird song assuring us that new life is imminent, to the hot and dry summer months when the evenings are filled with nighthawks, moths, bats, and owls, and then on to the fall colors—red in the understory shrubs and yellow in the needles of larch. We look forward to, and are comforted by, these kinds of predictable changes. Seasons also remind us that we have repeated chances to recoup, start over, reap the rewards of our labor, and then contemplate it all.

In contrast with the slow change associated with the appearance of different seasons, *ecological disturbance* represents a different kind of natural change that occurs not gradually and predictably, but instantaneously and unpredictably. To this point, I have been using the word "disturbance" rather casually, but ecological disturbance is

R. L. Hutto, *A Beautifully Burned Forest*,
https://doi.org/10.1007/978-3-032-03180-8_7

understood to be something rather specific. In ecology, a disturbance event is defined as a sudden change in the plant and animal community followed by a lengthy return to something like what occurred there before the disturbance took place. I need to expand on this idea because the only way to understand how so many plants and animals have come to depend on fire is through a basic understanding of naturally occurring disturbance events. A lot of things might disturb the status quo in nature. For example, conditions are disturbed when an elk rips off a piece of a plant, or when a fox captures a mouse, or when a rock rolls down a hill and crushes some plants. Each of those kinds of events constitute a disturbance in one sense of the word but, technically speaking, they are not the kinds of things ecologists formally classify as disturbance events. In ecology, only those things that cause a wholesale change in the plant and animal community within the disturbed area qualify as ecological disturbance. Thus, according to White and Pickett, disturbance is "...any relatively discrete event in time that disrupts ecosystem, community or population structure, and changes resources, availability of substratum, or the physical environment" [1]. Seth Reice views disturbance as a physical force that damages natural systems and removes organisms [2]. Platt and Connell write quite simply that disturbance is "...a discrete event that damages or kills residents on a site" [3]. Most definitions of disturbance (like those just presented) include value-laden terms like "disrupt" or "damage" or "destruction" or "devastation." There is an unintended consequence of using such terminology—those words imply that a green, mature forest condition is what ought to be! The negative connotation associated with those words could be avoided through a more judicious choice of words. For example, White and Pickett could just as well have said "alter ecosystem," Reice could have said "alters natural systems," and Platt and Connell could have left out the "damage" altogether. Definitions are important because they influence our subconscious and, boy, have we all been led toward a sense of destruction when it comes to the various forms of ecological disturbance.

Nevertheless, the *sudden death of individuals* over a fixed area is the key criterion of an ecologically significant disturbance event. Thus, by definition, ecological disturbance involves relatively severe events. Note that the death of individuals is free of the value-laden concept of damage or destruction. The greater the magnitude of the disturbance, the greater the area and number of individuals lost. Examples of disturbance events might include volcanic eruptions, blowdowns, hurricanes, tornados, floods, fires, landslides, and even treefall gaps. The area affected by a disturbance event varies widely—from a single tree that crashes in a tropical forest, to thousands of trees that are killed by a disturbance agent. A single tree being killed by a storm certainly changes the tree's life state rapidly, and it will take time for another individual to fill the same space, but in a fundamental sense that doesn't constitute an ecologically significant disturbance event because too few individuals are lost and the individual lost is usually replaced by another individual of the same species. There are exceptions, of course, and it is possible for the loss of a single individual tree to have significant ecological consequences when the death results in a succession of *different* species occupying the newly opened space (e.g., gap-phase dynamics in the tropics). When disturbance causes the death of entire stands of vegetation, however, we reach the more typical spatial scale for discussions of major disturbance events. I should note that relatively slow changes in community composition due to something like global warming are not considered to be disturbance events. Disturbance-mediated changes are characterized by nearly instantaneous changes in resources, substrate availability, and the physical environment.

Not all disturbance events are equal in terms of their effects, even the larger events. Ecologically speaking, relatively large, infrequent, severe forms of disturbance (LIDs) affect the *overstory* as well as the understory. In contrast, frequent, low-severity disturbance events move primarily (but not exclusively) through the understory. The more severe forms of disturbance are in a class by themselves because they set the stage for what happens next—the process of *succession*, where we observe not only different individuals coming to occupy the disturbed site, but individuals of different *species* that were not present before the site was disturbed. In plant ecology, the "...concepts of disturbance and succession are inextricably linked" [4], and you can't have succession without a relatively severe form of disturbance. A forested landscape consisting of numerous forest stands of different ages is not the result of numerous low-severity, understory fires—those kinds of fires do not initiate the process of dramatic forest succession.

Any ecology textbook will describe the process of succession as a predictable, linear sequence of change from early post-disturbance seral stages dominated by species that do well in open environments to later post-disturbance stages dominated by species that do well in more closed conditions (Fig. 7.1). I'm sure most everyone is familiar with the concept of plant succession following severe disturbance. It is not entirely clear why plant community composition changes as time proceeds following a disturbance event but, basically, disturbance creates a changed environment within which we first see species that are either super colonizers or fast growers. They then remain in place for a period before being outcompeted by more slowly growing but

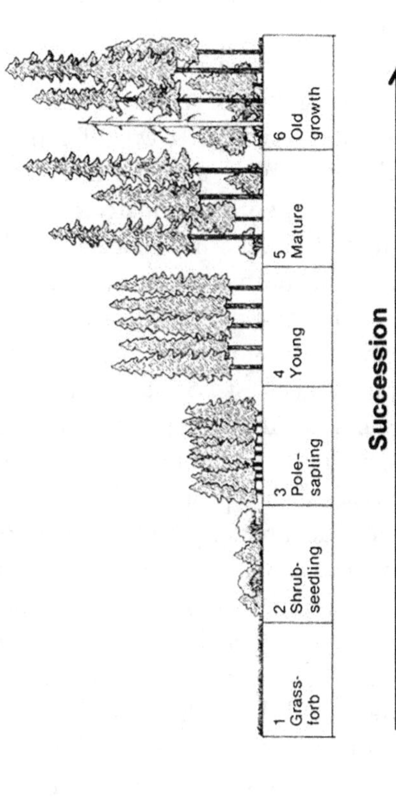

Fig. 7.1 A simple diagrammatic illustration of the process of succession following a disturbance event that kills and removes all the overstory plants. (Modified from Figure 30 in Thomas 1979)

competitively superior individuals of the same species that dominated the scene before the disturbance event occurred. Anyone can observe the largely predictable, unidirectional change in species composition after a disturbance event. The earliest stage of succession constitutes a relatively short period, and subsequent stages occupy successively longer periods until the process slows to a near steady state hundreds of years later.

Unfortunately, the textbook view of succession as a linear path toward some sort of climax condition (or end point) tends to draw attention toward species and conditions at the end point. This depiction, plus the fact that earlier stages of succession are very brief compared to later stages, fuels the thought that anything other than a relatively stable end-point condition is abnormal and constitutes a *disruption* or *disturbance* of the end-point condition. We need to get over this idea that succession is a linear process leading to a golden end point that best not be disturbed. Instead, we need to understand that different plant communities exist within a never-ending *circle* of time. Imagine the different stages of succession being positioned around a circle. That depiction would do a better job of implying constant change because environmental conditions always loop back to the earliest (or at least an earlier) post-disturbance point on the circle when yet another disturbance event occurs. The idea of constant change was captured well by Seth Reice when he proclaimed that "The normal state of communities and ecosystems is to be recovering from the last disturbance." Natural systems are so frequently disturbed that they rarely exist in a condition that one might consider a steady-state end point. In the northern Rocky Mountains, disturbance from severe fires is so common that it prompted botanists Habeck and Mutch to write, "anyone knowing forest successional patterns and species relationships can readily observe that a high percentage of the vegetation, within all forest zones, is at one stage or another of succession following past fires" [5]. I also find inspiration from Brandegee's 1899 report that in the Grand Teton area, "…fires have swept over the country so completely and persistently that scarcely any part has been entirely exempt from them…" [6] Basically, the forest landscape consists of numerous patches of different-aged stands and few, if any, of those stands are in what ecologists would call a steady-state climax condition. I would add that the near absence of climax vegetation conditions in severe-fire-dominated plant communities extends well beyond the Northern Rockies. If you want to be impressed with how little forested land in the West exists in an undisturbed "climax" condition, just look at the Capital Public Radio's California Wildfire History Map [7], which was generated from CalFire's database across a brief 100-year period—it shows that relatively little area in the entire state has escaped fire, and that the entire state seems to be a patchwork of burned areas in different pre-climax stages of succession.

Again, not all disturbance events are equally intense or severe. Large, infrequent, severe disturbance events (LIDs) that affect the overstory of a plant community are the most ecologically significant. This is because LIDs kill a lot of individuals, which sends things to the earliest point on the succession circle—a place where only the earliest-stage plant and animal species occur. The magnitude or extent of plant death or loss is great enough in a LID that things return to a point where *previously rare species* can get footholds early on after the disturbance event. Thus, the earliest stage after disturbance provides an open door for species that were not there before the disturbance event. This narrow window of opportunity associated with the earliest successional stages is when the magic happens—a patch of land that was in a later stage of succession becomes instantly transformed from that stage to the earliest stage. One can observe community composition in the disturbed area change rapidly before the process of succession continues to take it through a series of different communities toward something nearly identical to that which occurred just before the disturbance event happened. The turnover in species (the change in species composition) associated with succession makes disturbance an important ecological and conservation issue. The variety of life on Earth depends in no small part on natural disturbance events, which allow any one area to support different species through time, complementing the way a larger region supports different species at different locations within that region simply because physical conditions differ from one point in space to the next.

Thus, many species occur at the same time across a larger landscape, in part, because a series of independent disturbance events within that landscape create a kaleidoscopic "shifting mosaic." That is, different species occur in different places at the same time and all the species occur in one place but at different times. Think about this. In a disturbance-dependent plant community, every species is destined to occur in different places at different times across the larger landscape. A Black-backed Woodpecker might occur in one place this year and somewhere else the next. The same goes for species associated with old growth. It might seem like they will occur forever in the same place, but not if the old-growth patch gets a nice dose of disturbance; in that event, the species occurrence pattern will shift to encompass another place that just turned from a mature stage into an old-growth stage of succession. An intact, disturbance-dependent conifer forest will have all its component species, but never in the same places through time—that's the idea behind a shifting mosaic, and that is what land managers need to think about when it comes to maintaining the integrity of a disturbance-dependent community.

I remember when Mount Saint Helens erupted in 1980. I was walking home from campus in the afternoon and by then the air in Missoula was filled with ash. Enough ash settled on car windshields that people were scribbling things like "Help me!" on them. That rare, but natural disturbance event stimulated a lot of research, of course, and it led Jerry Franklin and co-authors to add significantly to our thinking about disturbance by emphasizing that disturbance is also an *editing process*; it's selective removal of some elements, and selective retention of other elements [8]. A lot of the plant and animal life following the eruption emerged from around or under what little organic material remained. The editing process (what remained following disturbance) had a good deal of unpredictability from one affected place to the next because severity of the disturbance varied from one place to the next. The species that occurred immediately after disturbance (and maybe even the trajectories of species composition during the succession process) were, therefore, highly variable as well. The precise rules of species removal and retention depend on the timing and severity of the disturbance, and no two disturbance events will be exactly alike. This concept is profound because it means that both the severity of the disturbance and the composition and structure of a pre-existing plant community influence the development of the future plant community. There is variability and unpredictability associated with the "start points" in succession due to variation in both the pre-disturbance vegetation condition and the kind of disturbance a place experiences. Therefore, a nearly infinite number of possible successional pathways are possible, and the range of community composition possibilities proceed from extremely diverse early on to much more similar later in succession.

We now know that there is considerable variation in start points and pathways associated with succession and that the succession process can continue for hundreds of years. Nevertheless, it's not uncommon to hear land managers and scientists express a fear that any unusual vegetation trajectory following severe fire represents some kind of vegetation "type conversion." For example, after the recent Rim Fire in and around Yosemite National Park, a botanist was quoted in a Los Angeles Times article [9] saying that now, after the fire, "if we don't intervene, it will convert to brush" [9]. Indeed, climate change is making the prospect of permanent type conversion more and more likely for low-elevation conifer forests. However, conversion to "brush" can also be part of a perfectly normal path of plant succession following severe fire. In the Sierras of California, forests at severe risk of type conversion comprise less than 10% of the forested area [10]. Thus, in most conifer forest types, true type conversion (where the original vegetation is permanently lost to something entirely new) is still very rare; most of the observable vegetation change following fire is merely a reflection of natural succession taking place, and successional pathways are varied. In most cases we'll have to wait decades to see if recently documented decreases in seedling survival due to climate-induced, hotter, and drier conditions fail to allow for a return to the forest condition present before the fire.

Severe fire disturbance is a naturally occurring phenomenon that has existed throughout most western conifer forest types for millennia. Moreover, the earliest stage of succession following severe fire in conifer forests is different from the earliest stage of succession following any other kind of severe disturbance event. The earliest stage of succession following fire includes a legacy of standing-dead trees—not a few snags per hectare (such as what might occur in a healthy mature forest), but hundreds of mature-sized, blackened snags per hectare. The legacy of what remains after a disturbance event is key to understanding why conifer forest fire disturbance events cannot be replicated by any other form of disturbance—none. You might be able to walk into a forest after a windstorm or a volcanic eruption and see hundreds of flattened trees per ha, or into a forest that has hundreds of hectares of standing dead trees from beetle kill or hurricane, but only severe fire disturbance leaves behind *hundreds of burned and blackened, standing dead trees per hectare*. The legacy of standing-dead trees following severe fire creates a picture of forest succession (Fig. 7.2) not typically captured in most textbook depictions of forest succession, where the first stage after disturbance is usually represented as a homogeneous moonscape devoid of dead trees. The standing-dead trees following fire also persist for decades, making the nature of succession following fire disturbance in conifer forests unlike the nature of succession following any other kind of forest disturbance. There are plant and animal species unique to each stage of succession following any kind of severe disturbance, of course, but many of the species attracted to a forest disturbed by fire are not attracted to conditions following any other kind of severe disturbance. Only fire can create blackened, standing-dead trees, and only fire disturbance can attract many of the species that occur during the earliest stage of postfire succession (e.g., morel mushroom, Bicknell's geranium, jewel beetle, Black-backed Woodpecker, Mountain Bluebird). This is important, so I'll repeat myself—no other form of disturbance (e.g., timber harvest, blowdown, beetle kill) can replicate

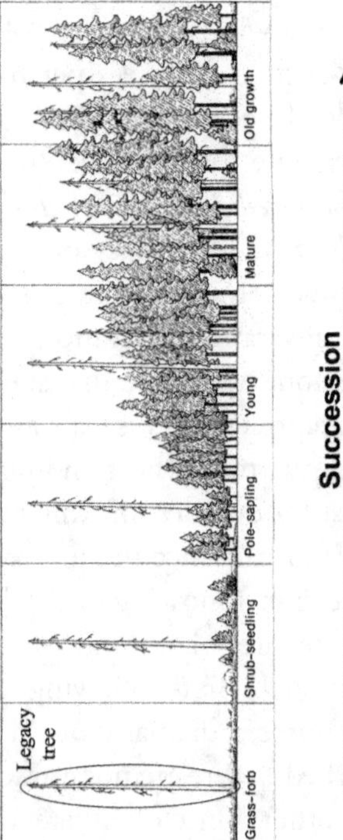

Fig. 7.2 Textbook depictions of forest succession rarely provide an accurate picture of succession following severe fire because they fail to include that all-important legacy of blackened, standing-dead trees, as depicted here. (Modified from Figure 39 in Thomas 1979)

the conditions that severe fire creates as a start point in succession because no other form of disturbance can create thousands of hectares of blackened, standing-dead trees. Therefore, no other form of disturbance supports the same successional series of plant and animal species. We can try to "mimic" nature through mechanical treatments, but our best attempts can't replicate the severely burned postfire conditions that historically occurred in most mixed-conifer forests [11]. Indeed, my earliest fire research showed that the bird communities in fire-disturbed and artificially disturbed forests are not anything alike [12].

Do severe, crown-fire disturbance events in other vegetation types around the world create conditions wherein a brand-new set of bird species colonize for a brief period before succession moves things along and they disappear again? Not to a degree that is even close to the kind of dramatic change in bird species composition I've been highlighting. A marked change in bird species following severe fire appears to be unique to mixed-conifer forest systems of western North America. Although there are certainly bird species that suddenly appear after fire in, say, eucalypt forests of Australia or chaparral shrublands of California, most of them are the same species that were present before fire, and none of them is as restricted to its respective postfire environment as the Black-backed Woodpecker is to a recently burned mixed-conifer forest environment. Moreover, the sheer number of species that suddenly appear seemingly out of nowhere following fire is nowhere near as astounding as it is in mixed-conifer forest. I suspect this is because other ecosystems that burn severely are dominated by plant species that remain in place and resprout from epicormic shoots and root crowns following severe fire disturbance, so the systems rapidly come to resemble the plant community that existed prior to the fire disturbance. Consequently, they retain the same species that were present before fire rather than attracting a brand-new set of species following fire. Pure ponderosa pine forests of the Southwest, which experience predominantly low-severity fires, also fail to produce a pronounced successional transformation following fire, so the bird species composition does not change markedly following fire in that forest type either.

Even though constant, disturbance-mediated change is etched into mountains of the West, it's still hard for us to accept the idea that things are always in a state of flux. Human thought is dominated by the view that some endpoint in succession is what ought to exist, and that carries with it the potential for well intentioned, but misguided management. For example, chief of the Forest Service, Mike Dombeck, wrote after the fires of 2000, "Restoring our forests to health will take more than a just a few years" [13]. By implication, recently burned forests are unhealthy, and we need to *restore* them to some state other than the wonderful state they're in after disturbance. Didn't those forest systems just get "restored" by those fires? Don't we need all stages of succession following disturbance? It depends on one's perspective and one's appreciation of disturbance as a natural process. The same sentiment exists today, as evidenced by Agriculture Secretary Tom Vilsack's 23 June 2022 memorandum to the USDA Forest Service, where he directed Chief Randy Moore to develop plans for "…accelerating reforestation and boosting nursery capacity to grow more tree seedlings for postfire recovery."

It is unfortunate that the word "disturbance" carries with it the implication that a normal order has been upset. But our inability to accept change is not simply a product of accidentally using a misleading word—we use that word intentionally. We have a love affair with lack of change even though homogeneity and stasis are illusory in nearly all ecosystems. A "predictable, equilibrium, one-and-only-one-climax-type-is-what-really-belongs-here" attitude follows from that love affair. That attitude, in turn, has influenced management policies. When a natural disturbance event occurs, land managers can't seem to wait to get back to what was. They want to prevent change and to "restore" forests to climax conditions as rapidly as possible should a pesky disturbance event sneak in. Nowadays, the most widely used new word in forest management is "resilience," or the ability of the forest to return to where it started before a disturbance event. Accordingly, our primary management goal these days is to make sure forests are "resilient" to fire and climate change and whatever else might move the system (even temporarily, it seems) from a timber-producing hunk of land. Most forests throughout the West are already plenty resilient, however; they just don't return to sawlog conditions as rapidly as managers would like. Land managers do what they can to speed the return to a mature forest (even to an extent that some forest managers plant trees and allow the use of herbicides) but speeding the successional process only operates to the detriment of all the species that depend on the early stages of forest succession. You now know what some of those species those are.

The big point here is that disturbance events are an integral part of most ecological systems. In fact, dynamism (as disturbance maybe should be referred to because even the word disturbance itself carries a negative connotation) is not only natural but necessary, as necessary as the sun itself to many species. The task of encouraging a more positive attitude about fire is really one of highlighting the role of disturbance in natural ecological systems. As Rogers [14] said so elegantly 25 years ago, "Just as ecologists and managers should strive to incorporate human values into management decisions, *humans in general will need to gain a better understanding of ecology*" (my italics) so that they can participate knowledgably in public land management and thus become part of a solution that strives to work with nature [14].

When a severe fire burns through a forest system, the place is transformed overnight from a land of green to a land dressed in black. The sudden change is jarring, and we never seem be prepared for, or accepting of, such change. Nevertheless, once we understand that abrupt change is a perfectly natural part of any ecological disturbance event, the whole disturbance cycle is as comforting and predictable as the seasonal changes we see annually. Not only that, we look forward to being periodically rewarded every time with that brief but magical forest transformation.

References

1. White PS, Pickett STA (1985) Natural disturbance and patch dynamics. In: Pickett STA, White PS (eds) Natural disturbance and patch dynamics. Academic, San Diego, pp 3–13, p 7
2. Reice SR (1994) Nonequilibrium determinants of biological community structure. Am Sci 82:424–435
3. Platt WJ, Connell JH (2003) Natural disturbances and directional replacement of species. Ecol Monogr 73:507–522
4. Johnson EA, Miyanishi K (2007) Plant disturbance ecology: the process and the response. Academic, Burlington
5. Habeck JR, Mutch RW (1973) Fire dependent forests in the northern Rocky Mountains. Quat Res 3:408–424
6. Loope LL, Gruell GE (1973) The ecological role of fire in the Jackson Hole area, northwestern Wyoming. Quat Res 3:425–443
7. https://projects.capradio.org/california-fire-history/#6/38.58/-121.49. Accessed 1 July 2025

8. Franklin JF, Lindenmayer D, MacMahon JA, McKee A, Magnuson J, Perry DA, Waide R, Foster D (2000) Threads of continuity. Conserv Biol Pract 1:8–16
9. Boxall B (2013) Rim fire's effects likely to last for decades to come. Los Angeles Times, Los Angeles
10. Hill AP, Nolan CJ, Hemes KS, Cambron TW, Field CB (2023) Low-elevation conifers in California's Sierra Nevada are out of equilibrium with climate. PNAS Nexus 2. https://doi.org/10.1093/pnasnexus/pgad004
11. Frank GS, Betts MG, Kroll AJ, Verschuyl J, Rivers JW, Swanson ME, Krawchuk MA (2025) Distinct bird assemblages emerge after fire versus forest harvest but converge with early seral forest development. Ecol Appl 35:e70032
12. Hutto RL (1995) Composition of bird communities following stand-replacement fires in northern Rocky Mountain (U.S.A.) conifer forests. Conserv Biol 9:1041–1058
13. Missoulian, 4 September 2000
14. Rogers P (1996) Disturbance ecology and forest management: a review of the literature. USDA For Serv Gen Tech Rep. INT-GTR-336:1-16

8

A Golden Opportunity

As luck would have it, my own fire research program received a significant boost when the Black Mountain Fire of 2003 burned through 3000 ha of forested land just west of Missoula. It started as a lightning-caused fire that burned for a week or two before 100-km/h August winds blew the fire into an inferno, chasing families from their homes as trees became engulfed in flames, throwing firebrands and ash for miles. I remember the day well. My wife and I were enjoying a picnic up the Bitterroot Valley at Lake Como when the winds suddenly whipped up. Our sons were in tow, and I remember how much fun we had standing on the edge of the dam, as those 100-km/h head winds from across the lake held us up at an angle while we leaned out over the water. By the time we had driven back to Missoula, the fire had exploded to become the model of a fire for this vegetation type. The resulting mixed-severity fire was typical of fires that burn throughout mixed-conifer forests in the West, and it took what was basically a high-severity,

R. L. Hutto, *A Beautifully Burned Forest*,
https://doi.org/10.1007/978-3-032-03180-8_8

pull-firefighters-off-the-line kind of fire to end up with the beautiful mixed-severity result we received then and can see even today, 22 years later.

I was excited about having a new study area pop up so close to town, and my wife (then an interpretive specialist for the Lolo National Forest) was also excited about the education opportunity the fire would provide—these kinds of opportunities don't happen more than once in a lifetime. We were among those who pushed hard for the USFS to leave the burned forest alone, and the District Ranger decided to do just that. This turned out to be one of the best decisions a ranger could make following a severe fire. Thousands of adjacent O'Brien Creek residents, school kids, and other citizens visited this area to learn about how amazing this wildfire was. These education efforts probably contributed to Missoulians becoming more ecologically "Firewise" today than they would otherwise be. I also suspect that Missoulians are more ready than residents of any other city in the nation to not only survive the next wildfire that burns close to town, but to hold a parade in celebration of the event when it happens! Field assistants and I worked most every day during the ensuing dozen breeding seasons in this burn, and the proximity to town allowed me to visit the burn frequently enough that I could learn some amazing new things about the process of succession after that beautiful disturbance event. All major fires near towns should be immediately transformed into educational opportunities for students and citizens of all ages; that's the only way people's perception of a burned forest will change—let them witness the magic for themselves.

So, what was it that I discovered in the Black Mountain fire? I got a better understanding of how the Black-backed Woodpecker is not only relatively restricted to burned forests, but to the more severely burned patches within the burn and to a relatively narrow time frame of the first half-dozen years following fire. I determined this from bird survey data collected from hundreds of points scattered throughout every category of fire severity and every successional stage across a 10-year postfire period. The story grew to involve more species than just the Black-backed Woodpecker. I recruited Dave Patterson, a statistician in the Mathematics Department at UM, to help me uncover significant fire effects on birds. We published a paper showing that 30 species (more than half the bird species breeding there) were significantly more likely to be detected in some combination of fire severity and time-since-fire than within the surrounding mature, unburned, green-tree forests of the same type [1]. That is to say, a lot of species not only reach their peak abundances in burned rather than unburned forests, but the exact combination of fire severity and fire age a species prefers depends on the species in question. I am continually blown away by data showing that many species (e.g., American Three-toed Woodpecker, Mountain Bluebird, Tree Swallow) become increasingly more abundant with every incremental increase in fire severity, even to an extent that outshines the Black-backed Woodpecker.

I also came to understand that Black-backed Woodpeckers are particular about the kind of burned forest they find suitable. It takes an *intact, mature, densely stocked,* forest with *severely burned, thick-barked trees* like Douglas-fir, western larch, and ponderosa pine to achieve the greatest response by the Black-backed Woodpecker and its wood-boring beetle food source [2]. Vicki Saab, the only other person I know who made a career out of studying birds in burned forests, showed the same thing in several of her publications and even provided a model that allows one to predict where burned-forest conditions are most suitable for the Black-backed Woodpecker [3]. Those forest conditions are why I chose to title this book "A Beautifully Burned Forest"—not *all* blackened forests are biologically beautiful. It takes an intact, mature, mixed-conifer forest before fire to generate a biologically beautiful postfire response.

The Black-backed Woodpecker story is not about single-species management or about conservation of the Black-backed Woodpecker—it is a story about how that one species plus countless other plant and animal species act as *indicators* of critical disturbed-forest conditions that receive inadequate management attention. Those indicators tell us that the historical fire regimes across most forests in the Northern Rockies clearly included healthy doses of severe fire, and those infrequently created burned-forest conditions need to be maintained across the larger landscape. I discovered that the Black-backed Woodpecker is a species that's talking to us about something bigger than itself; you can hear the message if you listen carefully.

As I noted earlier, it may take years for some species (e.g., Williamson's Sapsucker, Lewis's Woodpecker, House Wren, and White-breasted Nuthatch) to reach peak postfire abundance before their numbers decline again years later. These kinds of results underscore how a species may be fire-adapted, or even fire-dependent but one would never know without waiting until its successional stage had arrived. You can't conduct a comprehensive fire effects study by comparing unburned forests with burned forests that are only 1–3 years old. It takes more than several years to create ideal conditions for species that need large trees to break apart or for bark to begin peeling away from tree boles. I also learned about how western larch seedlings and Williamson's Sapsucker thrive near the edges between severely burned and unburned forest patches. I was able to observe pine dropseed and Bicknell's geranium and fire moss and other plants that botanists hadn't seen there before the fire. The

Fig. 8.1 A carpet of lupine following severe fire

mass flowering show by lupine (Fig. 8.1) and fireweed and a concomitant increase in native bumblebee pollinators made the place so special that I wanted to keep everything secret so I could have it all to myself.

In addition to my primary research focus, I also learned to appreciate that close-by fires can provide astounding educational opportunities. Through a partnership between the University of Montana, the Lolo National Forest, and the Montana Natural History Center, more than a few thousand kids were bused to the burn for outdoor education about fire. The brochure for an existing USFS nature trail at the edge of the burn was completely re-written to highlight elements related to the fire's beneficial effects. A local high school science class was able to make use of their block schedule to get bused weekly to the burn in 15 min, collect data on plant response to fire severity for an hour, and get bused back in another 15 min. We conducted nature walks for nearby residents and for the public at large, which stimulated some of the most satisfying comments from participants that a teacher can receive—things like, "Wow, I had never known these things. I will forever look at forest fires differently!" I was able to meet the forest supervisor for coffee and then take as little as 30 min in the field to hear him say "Wow…we have to come back with some of the other line officers so they can see these things for themselves." Basically, severe fire worked its magic.

Just before I retired in 2014, I wanted to capitalize again on the gold mine of bird survey data we had collected in association with the Northern Region Landbird Monitoring Program that we hatched 20 years earlier. That program had come to house data from more than 20,000 points that were scattered across western Montana. Many of those points were, by chance alone, positioned in places that had burned a handful of years before the bird survey was conducted. By using the bird data from those points, I could pursue another avenue to look at fire effects across hundreds of fires that had burned during the previous 25 years in western Montana. I was sitting on what was undoubtedly the largest combined fire-related and landbird monitoring bird-survey database in the world. All I had to do is collect additional data from within undersampled areas that harbored some important combinations of fire severity, fire age, and logging history. I went to the National Geographic Society once again and proposed mapping (within each fire perimeter) the locations of any pre-fire or postfire logging activity that occurred just before or just after a given fire within the fire perimeter. I superimposed maps of fire severity along with maps of logging history (Fig. 8.2) and conducted targeted bird surveys within those forest conditions that were undersampled in my already existing database. With support from the Society, my last formal field season took place in 2015 when my two college-aged sons, Rusty and Paul (Fig. 8.3), helped conduct surveys in a variety of previously burned areas. About half of the 68 species that we detected were more abundant in burned forest than in unburned conifer forest and, once again, the Black-backed Woodpecker was detected nearly exclusively in severely and recently burned mixed-conifer forests. Even more telling, we found the woodpecker species to be affected strongly and negatively by both pre-fire and postfire tree harvesting. Apparently, the most compelling indicator species need severely burned forests, and they don't like heavy-handed management before or after fire.

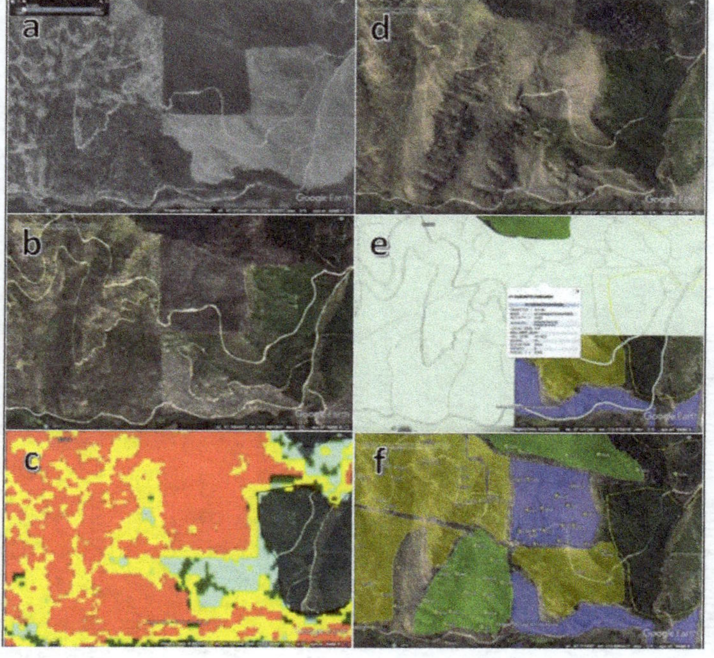

Fig. 8.2 An example series of photos showing the steps we took to classify potential survey locations in each fire (here, the 2003 Mineral-Primm fire). Before fire (**a**), we could see uncut and cut forest using Google Earth. After the fire burned (**b**), we could see that the uncut forest patch in the center burned severely, as also illustrated in the USFS fire severity map (**c**). That same forest patch was then salvage logged (**d**). The USFS management history data (**e**) then allowed us to create a final map (**f**) showing areas cut before fire (yellow), after fire (blue), and not cut before or after fire (green)

Fig. 8.3   My sons, Paul and Rusty, helping me collect bird survey data from older burned areas throughout western Montana in 2015

References

1. Hutto RL, Patterson DA (2016) Positive effects of fire on birds may appear only under narrow combinations of fire severity and time-since-fire. Int J Wildland Fire 25:1074–1085
2. Hutto RL (2006) Toward meaningful snag-management guidelines for postfire salvage logging in North American conifer forests. Conserv Biol 20:984–993
3. Latif QS, Saab VA, Haas JR, Dudley JG (2018) FIRE-BIRD: a GIS-based toolset for applying habitat suitability models to inform land management planning. U.S. Department of Agriculture, Forest Service, Rocky Mountain Research Station, Fort Collins

9

Interlude II

The gentle, soft "peer" notes of Mountain Bluebirds blanketed the soundscape when that first spring arrived following the Pattee Canyon fire, much as they did across dozens of other recently burned conifer forests I subsequently visited in different locations throughout the West. It is so special to see Mountain Bluebirds in a severely burned mixed-conifer forest because they would never have occurred there before the forest was transformed by fire (Fig. 9.1). The bluebird also serves as a reminder that there must have been plenty of old, dead trees standing in the forest before the fire burned. I know that because bluebirds use a disproportionately large number of cavities or holes that were excavated by woodpeckers in large snags long before the fire arrived. Because potential nest sites are rare early on following fire, bluebirds are not very picky; I found them nesting in holes created from broken branches and in holes from rocks rolling out of dirt banks in the road and even in holes created by burned out tree roots. I learned that the numbers

R. L. Hutto, *A Beautifully Burned Forest*, https://doi.org/10.1007/978-3-032-03180-8_9

Fig. 9.1 The Mountain Bluebird is one of several bird species that have an uncanny ability to find and colonize severely burned forests within a year following a severe fire. (Photo Milo Burcham)

of potential nest sites explode over the first few years following fire because trees fall, branches break, and new cavities are dug every year by woodpeckers and nuthatches—the primary cavity nesters. The bluebird is just one of many secondary cavity-nesting species, which include species that are unable to dig their own nest cavities but take advantage of nest sites excavated by the primary cavity nesters. Other secondary cavity nesters I've found in burned forests include species like the Northern Pygmy-Owl, Tree Swallow, Western Bluebird, House Wren, Mountain Chickadee, and even the Common Goldeneye—a duck! Speaking of Tree Swallows, the sky above a recently burned conifer forest will come alive with an airshow of swallows feeding on the wing as they bank and turn amid all the standing dead trees in the more severely burned portions of the forest.

Another iconic bird that loves severely burned forests is the Townsend's Solitaire. That bird certainly caught my attention early on because of its melodious song, which it often sang while on wing, flashing its golden wing patches as it flew. Some of its nest site locations were even more eye catching. These included little cave-like places where rocks rolled out of unstable dirt banks, and even in a place created by a rock rolling out of an uprooted tree mass that was uplifted when a tree blew over in the first few years following the fire (Fig. 9.2). While I'm on the topic of tree roots, the same kind of burned-out root holes in the ground that I saw bluebirds sometimes use as nest sites are commonly used by ground-nesting Dark-eyed Juncos as well. There were so many juncos nesting in the Black Mountain fire that one of my graduate students, Bruce Robertson, was able to find and monitor more than 50 nests per year as part of his dissertation work. And after the fires of 2000 burned in the Bitterroot Valley, Kristina Smucker, another one of my graduate students, found the very first ground-nesting Flammulated Owl ever reported. It, too, was using a burned-out root hole as a nest location. Until you see these things for yourself, it's hard to appreciate the biological value of the unique *forest architecture* created by a severe fire. Standing-dead trees and their immediate surroundings provide a critical part of the postfire forest architecture needed by so many fire-dependent species (Fig. 9.3).

The standing dead trees are a constant presence, of course, and they stand as spires in the air, casting shadows on the ground. I can never stop gazing at the tops of the blackened trees. Each tree species has its unique branching pattern, and one can become quite good at identifying those blackened tree species from a distance. The other

Fig. 9.2 Townsend's Solitaire nest in a site created when the wind blew down and uprooted a standing-dead tree. Dead trees created by severe fire are valuable as sources of insect food and nest sites not only for the years after a fire when they are still standing, but even after they blow down

Fig. 9.3   Burned-out tree root holes can capture an unsuspecting hiker, but they also serve as nest sites for juncos, bluebirds, and even small owls

Fig. 9.4 Treetop cones of lodgepole pine. Several conifer species maintain cones high atop their blackened remains following a crown fire, and those cones serve as important sources of seed that contribute to recolonization from well *within* high-severity burn patches

thing I came to appreciate increasingly was how clusters of open cones on most conifers occur at the very tops of the blackened trees (Fig. 9.4). Most of those cones open in response to the heat of fire and continue to release viable seed long after (in many cases years after) fire kills the tree. By looking at the distribution of seedlings some years after fire, it becomes abundantly clear that the reforestation process is in no small part a product of re-seeding by standing dead trees situated well *within* a fire's perimeter, and not entirely from some unburned green-tree forest edge. Recent research even shows that closed cones of ponderosa pine, Douglas-fir, and lodgepole pine scorched by severe fire's heat can still release viable seeds (a condition the authors call conditional pyriscence) well within large high-severity fire patches [1]. High atop blackened trees, open cones of Douglas-fir and ponderosa pine also attract the attention of the Clark's Nutcracker and seed-eating finches like the Pine Siskin and Red Crossbill. The nutcracker has a reputation of dispersing and planting whitebark pine seeds at higher elevations, but I came to appreciate that the nutcracker also spends a good deal of time at lower elevations extracting seeds from ponderosa pinecones that have become available after the heat of fire causes even those cones to open slightly. The nutcracker is one of only two songbird species that has a pouch beneath its tongue within which it can store up to 150 seeds until it heads off to cache them in small groups on an exposed hillside somewhere. They have amazing spatial memories and can recover buried seeds months later. But because they never recover and consume every seed stored, they become amazing tree planters as well.

The scientist who discovered a lot about the seed-storing behavior of nutcrackers was Russell Balda. He set up a nutcracker experiment room at Northern Arizona University when I was a graduate student there. He scattered dozens of small, dirt-filled cups, rocks, and other items across the floor and let nutcrackers store some seeds in a small subset of the cups. He challenged his graduate students to watch and remember where birds stashed their seeds. He then brought both the birds and students back into the room a month or two later and compared how well they remembered where the seeds were stored. The birds outperformed the students by a wide margin. When Russ presented these amazing results at an ornithological conference, he excitedly emphasized that the birds even performed better than university graduate students. At that point, someone in the back of the room suggested that maybe the results reveal more about the quality of his graduate students than the spatial memory of nutcrackers! Anyway, I was working in the 1988 Lolo Fire near Missoula when I managed to capture my own lousy photograph of a nutcracker with its sublingual throat pouch chalk full of ponderosa pine seeds (Fig. 9.5). I know some people might doubt whether nutcrackers extract and disperse ponderosa pine seeds from open cones on burned trees, but it is not at all uncommon. The biology surrounding blackened, standing-dead trees in burned forests never ceases to amaze me.

Fig. 9.5   Clark's Nutcracker extracting seeds from cone on a blackened ponderosa pine tree. Note the bulging sublingual throat pouch containing pine seeds that have been extracted

Reference

1. Lopez MA, Kane JM, Greene DF (2025) Seed maturation and mortality patterns support non-serotinous conifer regeneration mechanism following high-severity fire. Fire Ecol 21:10

10

Reconstructing an Ecologically Relevant Past

There is concern these days about whether our forests are currently outside the range of natural variation and are experiencing types of fires they have never harbored before (i.e., whether fire regimes have departed from historical norms). Many believe that the way things used to burn might be good to understand, but those past fire regimes are no longer with us—things have changed due to the relatively recent influence of humans. Why should we still want to discern the kind of environment that existed in the past? The answer is that ecologically relevant past environments provide the backdrop against which existing plant and animal species evolved, so past environmental conditions are what we must maintain if we want extant species to persist. Because public land management agencies are legally responsible for maintaining the integrity of living systems, they certainly ought to understand whether current forest conditions and current fire behaviors are normal and within the historical range of natural variation or whether they are,

R. L. Hutto, *A Beautifully Burned Forest*,
https://doi.org/10.1007/978-3-032-03180-8_10

indeed, out of whack. If things are out of whack, then we might want to do something to "fix" the situation. So, how do agency personnel come to know whether the patterns of burning in a particular place have changed from historical norms? They rely on fire scientists. But how do fire scientists reconstruct the historical fire regime for any one place? I'll first tell you about the most common approach scientists have used, and then I'll introduce you to an underused but more ecologically oriented approach that can also yield insight into past environments.

Any effort to uncover a pattern of burning from the past will have to involve a particular period for establishing that history, but which period is appropriate? Since a regime requires a substantial block of time to reveal itself, the target should be an ecologically relevant historical period—a period that reflects the historical environment within which the animals and plants we see today evolved. The recent past is too tainted by "unnatural" events that do not reflect the historical conditions under which native species evolved and now require, so we must get back in time prior to when there was a heavy-handed, widespread, and unnatural human influence on the land. Otherwise, we might be managing the land to maintain what could very well be unnatural conditions. In doing so, we would risk losing the very plant and animal species that we're trying to maintain. With respect to fire cycles, if they occur on the order of dozens or even hundreds of years, then it would take a period of hundreds or even thousands of years into the past to uncover that natural pattern.

Traditionally, fire scientists have used fire scars on trees to reconstruct when fires burned in the past (Fig. 10.1). Fire scars occur on trees that were injured by fire at one or more points in time. What happens is that a fire injury kills cells in the outer cambium. After that, there is no longer cell

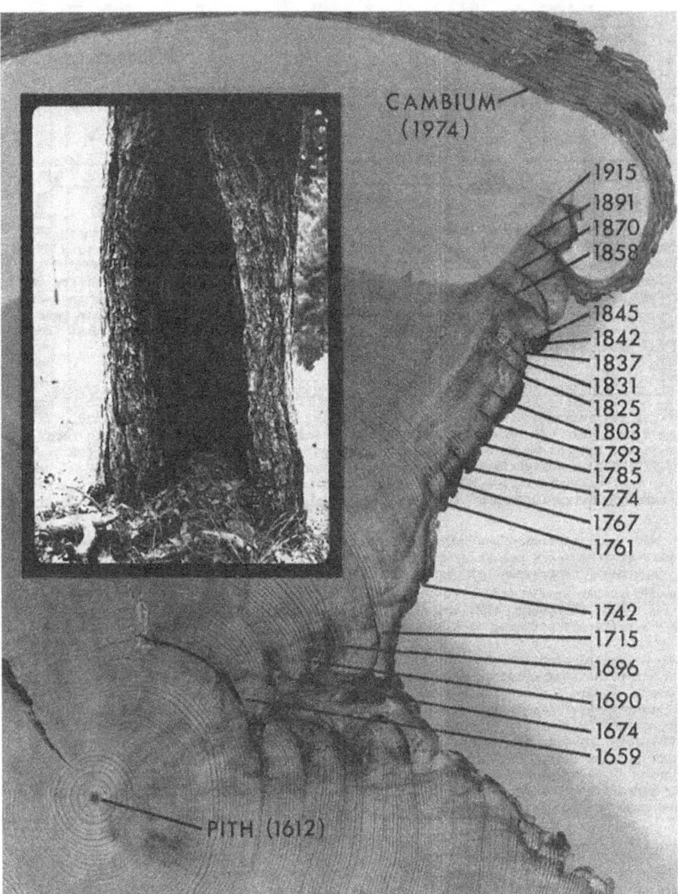

Fig. 10.1 Cross section of a tree with fire scars. The year associated with any one scar is determined by counting the number of tree rings between the pith and the ring with the scar. (From Figure 3 in Arno 1976)

growth in that spot, although adjacent cells begin to grow over the injured area and create a visible scar. Every fire that follows is then likely to re-injure the tree at the same sap-rich scar location, and adjacent living cell tissue will begin to grow over the injured area again. Fire scientists count the number of growth rings in an uninjured section of the tree to determine the tree age and then note the specific rings (years) in the scarred section that show evidence of having received an injury. Although exceptionally old trees can expose a thousand or more years of fire history, fire scar records from most trees generally allow one to go back only a few hundred years, so that aspect associated with fire-scar records can be problematic. Nonetheless, a telling story has emerged from these fire scar studies.

The fire scar records show that hundreds of years ago forests used to burn frequently (maybe every 6–10 years, on average, in the extreme southwestern US), but after the turn of the century they began to burn infrequently if at all. If you line up a bunch of fire-scar records from dozens of individual trees growing in one place, one atop the next, there are very few scars showing to the right of the twentieth century mark on the time axis—it's like a fire light switch was flipped off (Fig. 10.2). The human activity most frequently blamed for the sudden decrease in fire frequency is fire suppression, but that story is undoubtedly overblown because fire suppression efforts never really kicked into high gear and were never that effective until after the second World War. Instead, timber harvesting, grazing (through the removal of flammable grass cover by cattle and sheep), and a wetter climate are probably the influences that played more important roles in the abrupt decrease in fire frequency starting at the turn of the century.

Fig. 10.2 Each horizontal bar shows years with (in red) and without (in gray) a fire scar for a single tree. Note how frequent scarring on most trees ends abruptly around year 1900. (Modified from Figure 1 in Marlon et al. 2012)

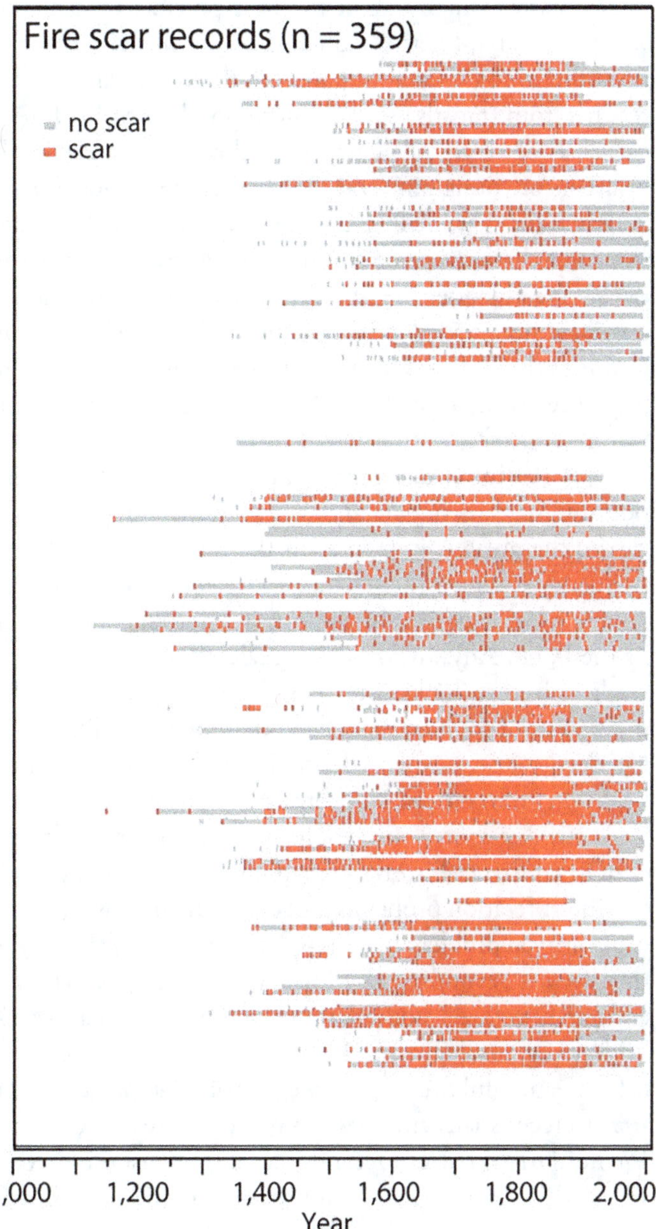

Fire scar records (n = 359)
no scar
scar
1,000
1,200
1,400
1,600
1,800
2,000
Year

No matter which one of the four possible explanations was responsible for the sudden decrease in fire frequency, one other thing became apparent to those studying fire scar records from (mostly) southwestern ponderosa pine forests: the absence of frequent fire also allowed increases in the numbers of young pine and Douglas-fir trees. Most fire scientists argue that the increase in numbers of smaller understory trees has now created a situation where the understory trees serve as ladder fuel should a fire occur, which leads to atypical (more severe) fire behavior. In this book, I will refer to this as the "forests are out of whack" story as a shorthand way of referring to the kind of fire-related structural change in forest species composition I just described. Because the story carries important management implications, it is important that we understand which potential explanation carries the greatest influence on the sudden change in fire frequency (or any other characteristic associated with a fire regime). I will address this issue more thoroughly at the end of the book. Anyway, on a timescale afforded by tree-ring studies, it sure looks like fires burned frequently (every 6–10 years or so, on average) and that forest conditions and fire occurrences changed dramatically from century at the beginning of the twentieth century.

As striking as tree-ring results may be, the use of fire scar data to construct historic fire regimes does not come without interpretation problems and does not come without the more critical problem of extrapolating results to forest types that lie well beyond the dry-forest types where the data were collected. Before I address those problems, remember the underlying problem associated with fire-scar data—ecologically meaningful fire regimes are probably *too far removed in time* to reconstruct from relatively recent fire scar evidence. We need to get back thousands rather than just hundreds of years. Relatively recent fire scar records are heavily influenced by relatively recent human activity (e.g., burning, grazing, introduction of exotics, urbanization, fire suppression), so they may not be able to generate an evolutionarily meaningful baseline.

In addition to the sample period being limited, there are some tricky data interpretation problems associated with fire scar analyses. For example, when it comes to calculating the average inter-fire interval, practitioners generally exclude the above-average number of years it takes for a tree to receive its first scar. The justification for ignoring the time interval between a tree's first year of age and its first fire injury is that it takes many years for a tree to grow large enough to receive an injury but not die. The initial injury also takes a relatively infrequent locally severe event. After that first injury, every subsequent fire will re-injure the tree in the same spot, creating a meaningful record of fire occurrence from that point on. Whatever the justification, exclusion of the relatively long year-to-first-scar interval will lead to an overestimation of fire frequency by some unknown amount. Another data interpretation problem is that fire ecologists create a fire history from a "composite" index, which is a combined picture of recorded fires across all trees sampled within an area. The problem is that many fires are very, very small, and may injure only a few trees in a large area. By throwing fire-scar dates from a spatially extensive pool of trees into one pot, one can be fooled into believing that fires burned more extensively and more frequently across the *whole* sample area than they really did.

Finally, and most importantly, many scientists who use fire-scar data focus way too much on the "average" fire return interval and not enough on the range of natural variation in return interval. I once went on a show-me tour about how "out of whack" the mixed-conifer forests are in my part of the woods. The leader noted that the forest we were standing in was way overgrown and in unhealthy condition; you know, it's the same story you hear on public presentations throughout the West. His estimate of the average number of years between fires there was around 12 so, he argued, the place must have missed 4 fire cycles because it hadn't burned there for nearly 50 years. The only way he could have been right in arguing that the forest was out of whack, however, is if there were no *variation* around the mean of 12 years. Was the mean 12 and the standard deviation zero? People that claim a place has missed several fire cycles drive me crazy because they ignore the fact that means carry standard deviations. We need to look at the variation around the mean before we can conclude that a fire-free interval constitutes an extreme outlier. A mean means nothing without its associated variation. Maybe in the past that place we were standing in had some fire-free intervals that were 80 years long and other intervals that were 3 years long. Unless we can see the entire distribution of the number of years associated with a sample of, say, 100 consecutive fire-free intervals from the past, we can't make an informed decision about whether things are out of whack or not. We've got to have a good fire history, complete with statistical distributions, to make judgements about whether the current situation is unusual. Using the average fire return interval as a basis for claiming that a forest is out of whack is problematic at best and dead wrong at worst. All told, the shortcomings of fire history information "collectively limit the potential of fire history data to provide ecologically meaningful guidelines for fire and land management planning," according to Stephen Pyne [1].

Researchers who have paid close attention to these possible biases have conducted additional fire-scar and historical reconstruction studies, and these studies tell us something quite different from the usual "frequent, low-severity" fire story. These more recent studies suggest that most forests across the western US (those falling within the mixed-conifer and higher elevation zones) were historically structured not by low-severity fires exclusively, but also by infrequent high-severity fires yielding mixed-severity effects [2]. Low-severity, understory fires, which many fire scientists believe dominated conifer forests historically cannot yield mixed-severity effects. Only high-severity fires can yield a patchwork of mixed-severity effects across space and time. Even low-elevation ponderosa pine forests are typified by an unspecified amount of mixed-severity fire. This is precisely what I argued earlier; all fires experience all severities—it's simply the proportions of the different severities that differ among fire regimes.

On top of possible data interpretation issues lies a much larger problem with fire scar studies. Fire scientists tend to *extrapolate* their research results to well beyond where the data apply. For obvious reasons, fire scar studies commonly target old-growth stands or stands containing trees with abundant fire scars. Fire-scar studies are, therefore, necessarily biased toward places where surface fires scar, but do not kill, trees and that, in turn, automatically biases such studies toward finding evidence of a low-severity fire regime. The results may very well apply to the place of study, but the biased sampling means one must be extra careful not to extrapolate results beyond the place where the data were collected. Nevertheless, the most frequently heard and most widespread story that has emerged from fire-scar studies is that, prior to the twentieth century, *western conifer forests* used to burn, on average, every 6–10 years. After the turn of the century, fire suppression, grazing, and other human activities eliminated fire and allowed understory fuels to build up to the point that things are now "out of whack" so that when a fire does return it burns more severely than ever. Even though this "forests are out of whack" story has been extrapolated to places and vegetation types that lie well beyond where the data were collected, it is now so widely accepted and so widespread that if you haven't heard it, you probably live under a rock! Wherever you go in the West, whatever the conifer forest type, that's the story you hear.

So, how did the "forests are out of whack" story ever get the traction needed to move well beyond the forest locations and types where that story applies? There was, at first, a small seed planted in the early work of some ecologists who wrote that forest fires are natural and essential, and that any effort to prevent or suppress fire in our wildlands would be shortsighted and destined to lead to out-of-whack forests and out-of-whack fires down the road. The earliest telling of this "forests are out of whack" story can probably be attributed to Harold Weaver, a fire ecologist who wrote in 1943 that "Because of these ecological changes [understory growth due to overprotection], which are continuing to take place, the fire hazard has increased tremendously. Fires, when they do occur, are exceedingly hot and destructive and are turning extensive areas of forest into brush" [3]. Weaver also wrote later about huge changes wrought by cattle grazing and fire exclusion. Charles Cooper also argued in 1960 that the most conspicuous change in the Flagstaff area since white man settled was an "…increase in area and density of pine reproduction stands" [4]. It is fascinating, however, that the extrapolation of research results to beyond where they apply is already evident in Weaver's early work. Back then, Arthur Brown noted in a comment appended to Weaver's paper that "Overstocked and stagnating stands seem so far from typical of the region for which he speaks [outside Arizona] that one wonders if Mr. Weaver is not generalizing too much from a single area."

Those early studies set the stage for important research that followed, and the subsequent research served to ignite a firestorm of attention and even a subsequent funding stream from the US Congress. The key studies were conducted by Southwest scientists who also showed that frequent, low-severity fires are largely absent now, but used to be the norm. More specifically, Wally Covington, who was on the Forestry faculty at Northern Arizona University, began looking at the structure of relatively small research plots across a period that was as extensive as his historical data would allow. Given he was working in a disturbance-based system, it should come as no surprise that the condition of any of the forest stands at the time of his studies were nothing like they were decades before; after all, change is the norm, right? So, why was the fact that Covington found a changed forest condition so noteworthy? It wasn't just one plot, but many plots that showed not cyclical, but unidirectional change from open-grown ponderosa pine forests to pine forests cluttered with an unnaturally dense understory. That seemed to be unusual enough for him to believe that the change was widespread and somehow connected, and that belief necessarily fueled the idea that we now have a widespread problem that needs to be "fixed."

This time around, the tendency to extrapolate results to beyond where they apply is still evident. The titles of Covington's published research began to expand from something that started with a local or regional flavor (e.g., "Post settlement changes in natural fire regimes and forest structure: ecological restoration of old-growth ponderosa pine forests" by Covington and Moore in 1994, or "Southwestern ponderosa pine forest structure: changes since Euro-American settlement" by Covington and Moore in 1994) to something involving forests of the entire US (e.g., "Helping western forests heal: the prognosis is poor for US forest ecosystems" by Covington in 2000)! Nowhere is the extrapolation and lack-of-nuance problem more serious than when scientists like Covington claim that dry forest ecosystems of the entire *West* are in "widespread collapse." Covington even accused those who think differently as either unethical or, at the very least, fuzzy thinkers [5]. Fortunately, others (e.g., Gutsell et al. [6]) have called him out on this.

Even as recently as 2012, Covington reiterated his results on National Public Radio when he said "Eighteen years ago, these pine forests had 50 times the number of trees that would have been here naturally. That's because for most of the 20th century, the Forest Service put out almost every fire. That allowed small trees to grow like weeds. In a fire, they become ladders that carry flames up into the taller trees and kill them [in a crown fire]" [7]. Sound familiar?

It is safe to say that most people today believe (as expressed by the General Accounting Office in 1999) that "the most extensive and serious problem related to health of national forests in the interior West is the over-accumulation of vegetation which has caused an increasing number of large, intense, uncontrollable, and catastrophically destructive wildfires." The story is so widely accepted that multiple forms of legislation have since been drafted to address the problem. For example, the Healthy Forests Restoration Act 2003, was justified because "…the forests and rangelands of the West have become unnaturally dense, and ecosystem health has suffered significantly." The Act says catastrophic fires are caused by deteriorating forest and rangeland health, and it provides for expedited environmental review to, in part, conduct fuels reduction to "address threats to forest and rangeland health, including catastrophic wildfire, across the landscape." Recently, President Biden also signed the Infrastructure Investment and Jobs Act of 2021–2022, which includes hundreds of millions of dollars to address the perceived threat of catastrophic wildfires by expediting thinning and prescribed burning. God only knows what the current Trump administration is going to do. Underneath it all, the belief that our forests are "unhealthy" or "out of whack" is bolstered by people's deep-seated inability to accept disturbance (or any kind of change), even when the change is natural.

Given the unnatural fuel buildup due to fire suppression (and, more importantly, grazing and timber harvesting), we might have a fire-related problem in the pure ponderosa pine forests of the Southwest, where fire size, frequency, and severity seem to have moved outside historical norms. For example, in a 2012 NPR interview, Swetnam stated that fires are now burning bigger and hotter. They're not just damaging forests—they're wiping them out. Swetnam goes on to say "Now the fire behaviors are just off the charts…I mean, they are extraordinary. Actually, I think in some cases, there's fire behavior that probably these forests haven't seen in millennia or maybe even tens of thousands of years." That story may very well be true for a small subset of southwestern forests, but I wish to emphasize here that *it is not true for most mixed-conifer forests throughout the West.* Nevertheless, forest managers beyond the Southwest continually apply that "out of whack" story (incorrectly) to their own forests and even to chaparral shrublands. One can read or hear most anywhere outside the Southwest that, due to excessive fuels, the fire sizes, frequencies, and severities we are witnessing are way beyond their historical ranges of variation. Even though forest managers are careful to acknowledge that "no two forests are alike," they always proceed with a decision to make their own forest more resilient to disturbance by thinning and burning!

Again, the main problem with a message that forest fuels are out of whack and fires are now burning unnaturally is that the veracity of both claims depends on the forest type and place. The claim that there is unnatural fuel build-up may be true for some vegetation types and in some places (e.g., ponderosa pine-dominated dry forests in the Southwest), and most scientists agree with Swetnam and others on that. The claim is certainly false for other vegetation types and places (e.g., higher-elevation lodgepole pine and spruce-fir forests), and everyone agrees on that as well. So, what's the problem? The problem is that *most* western forests lie somewhere in between those extremes, and while unusual change in forest structure may be perceptible in some of those forests, most still have compositions and structures that are well within the historical range of natural variation [8].

Clearly, the story that all our forests are "out of whack" needs to be reined in. In fact, the open, dry ponderosa pine forests that most would agree are closest to being outside the range of natural variation cover as little as perhaps 15% of western forested land. A recent estimate for forests of the Sierra Nevada labeled only 9% of the forested area to be so far out of whack with current climate that it is now deemed to be at "severe risk" of type conversion [9]. Still, it's not uncommon for forest professionals in Montana to tell me that fuels have built up unnaturally in our *mixed-conifer forests*, where the vegetation density hasn't changed perceptibly for centuries, or where a conifer understory has grown in a perfectly natural way as plant succession continues following some past natural disturbance event. Indeed, the need for forest "restoration" is being applied uncritically across Rocky Mountain forests and is now embedded in places like the Healthy Forests Restoration Act. Even if a mixed-conifer forest were slightly denser than the historical average, that does not mean it isn't perfectly capable of responding naturally to a severe fire when fire does return. Moreover, a natural reforestation response can take multiple dozens of years, so one cannot claim, after a period of 5–10 years, that a given forest is now threatened with "type conversion" to something other than the forest type that was there before disturbance. Outside of modeling successional pathways, don't you think one would have to wait 50–100 years before concluding such, especially given our understanding that successional pathways are infinitely variable?

I cannot emphasize the following point enough: except for a small fraction of western forests that include low-elevation, ponderosa pine forest types (mostly in the Southwest), the idea that years of suppression and timber harvest and grazing have created out-of-whack conditions is simply untrue. Most forests are well within their historical ranges of natural variation and fire behavior has not departed significantly from historical norms [10].

Fortunately, fire-scar studies are not the only way to learn about past fire environments. There are alternative methods scientists use to establish historical fire regimes, including historical ecology, the age structure of trees in a site, patterns in lake sedimentation rates, pollen or charcoal deposit analyses, and other palaeoecological proxies. Even though these methods may not be sensitive enough to pick up the frequent, less severe fire events, they can be used to infer severe-fire frequency over longer time periods (hundreds or even thousands of years). Not surprisingly, results from *every single one* of these approaches reveal that severe fires have occurred infrequently across most forest types in the West as far back as any of these methods can take us. The presence and potential importance of these severe-fire events is, quite simply, not captured in fire-scar studies.

In addition to all these historical methods, there is an even more powerful method we can use to reconstruct fire history and to uncover historically important fire regimes. The method takes us back to a period well clear of any unnatural human influence—back to the environmental conditions that steadily etched their signature on to the organisms we see today. The method involves understanding how natural selection works to create evidence of the kind of environments that sustain extant plant and animal species. It is the most underappreciated and underused method, even though it can yield a solid ecological understanding of which fire regime was historically important in any given area or forest type. It is also the method I highlight in the next chapter.

References

1. Pyne SJ, Andrews PL, Laven RD (1996) Introduction to wildland fire. Wiley, New York
2. Baker WL, Williams MA (2018) Land surveys show regional variability of historical fire regimes and dry forest structure of the western United States. Ecol Appl 28:284–290
3. Weaver H (1943) Fire as an ecological and silvicultural factor in the ponderosa pine region of the Pacific Slope. J For 41:7–14
4. Cooper CF (1960) Changes in vegetation, structure, and growth of southwestern pine forests since white settlement. Ecol Monogr 30:129–164
5. Covington WW (2003) Restoring ecosystem health in frequent-fire forests of the American West. Ecol Restor 21:7–11
6. Gutsell SL, Johnson EA, Miyanishi K, Keeley JE, Dickinson M, Bridge SRJ (2001) Varied ecosystems need different fire protection. Nature 409:977–977

7. http://www.npr.org/2012/08/24/159374096/is-it-too-late-to-defuse-the-danger-of-megafires. Accessed 1 July 2025
8. Baker WL, Hanson CT, Williams MA, DellaSala DA (2023) Countering omitted evidence of variable historical forests and fire regime in western USA dry forests: the low-severity-fire model rejected. Fire 6:146
9. Hill AP, Nolan CJ, Hemes KS, Cambron TW, Field CB (2023) Low-elevation conifers in California's Sierra Nevada are out of equilibrium with climate. PNAS Nexus 2:pgad004
10. DellaSala DA, Hanson CT, Baker WL, Hutto RL, Halsey RW, Odion DC, Berry LE, Abrams R, Heneberg P, Sitters H (2015) Flight of the phoenix: coexisting with mixed-severity fires. In: DellaSala DA, Hanson CT (eds) The ecological importance of mixed-severity fires: nature's phoenix. Elsevier, Amsterdam, pp 372–396

11

Adaptations Can Indicate Fire Regimes

I told you how, for me, evolution by natural selection was the most powerful concept I learned in college. Let's now use that concept to gain insight into past fire regimes. Imagine, if you will, a small owl that hunts by night. During the day, a resting owl is visible to daytime predators such as larger owls, hawks, and mammalian carnivores. Next, appreciate that there is, and always will be, variation in plumage coloration among individual owls of the same species. This is a necessary consequence of the independent assortment of, and minor mutations in, chromosomes that made up the eggs and sperm of their parents—no two owls have precisely the same genetic makeup, so no two owls look alike. Given this variation in plumage coloration, some individuals will automatically blend into their background and fare better than others when it comes to avoiding predation. The individuals that fare better in avoiding predation will, of course, leave more offspring than those that were not quite as well hidden. Over many generations, the

R. L. Hutto, *A Beautifully Burned Forest*,
https://doi.org/10.1007/978-3-032-03180-8_11

plumage coloration of all individuals will necessarily become more and more cryptic. This is the way *natural selection* works. It favors individuals that, through no conscious effort on their part, happen to have a genetically determined trait that allows them to do better than individuals of the same species that lack that level of trait development. In the future, due to the ever-present variation among individuals, the proportion of individuals that carry the most effective versions of the same trait will increase. The power of natural selection is why owls of all species have evolved to look so

Fig. 11.1 The plumage of this screech-owl illustrates how natural selection has resulted in the evolution of a plumage color that enables the owl to become cryptic in its native environment. Note that an innate behavior (to squint eyes and raise ear tufts) has also evolved. (Photo Steve Irvine)

amazingly cryptic against the specific tree species within which they tend to sit all day (Fig. 11.1).

I noted earlier that Dobzhansky captured the profound power and importance of natural selection when he wrote, "nothing in biology makes sense except in the light of evolution" [1]. Natural selection is the fundamental force that leads to the evolution of biological patterns we see today. Armed with this knowledge, we can envision a completely different approach to understanding fire history. We can work backwards from the traits of organisms that live in a burned area to uncover the historically important fire regime that must have promoted the evolution of those traits. This requires an understanding of the effects of fire frequency, severity, patchiness, extent, and timing on the occurrence and success of target organisms in the vegetation type of interest. Even though it's nearly impossible to design an experiment that would allow one to tease apart the effects of even one (let alone various combinations) of those fire variables on the occurrence or success of a target species, there are a lot of very convincing correlative studies. Some species are good "indicators" of certain postfire conditions because they are either restricted to those conditions (and that information emerges from studies of the distributions of plants and animals), or they perform relatively well only in those specific postfire conditions. An organism might perform well in one environment and not in another because it has evolved specific traits (e.g., behaviors, physical appearance) that allow it to have high reproductive or survival success in one environment and not in others. Those traits can be viewed as *"adaptations,"* which I would like to focus on here.

An adaptation is a trait that, on average, allows the individuals who carry it to do better than those who don't carry it in the environment within which it evolved. The evolution of an adaptation takes numerous breeding events because there is little chance that, in a single breeding season, individuals carrying a certain trait will always leave more offspring than individuals without that trait. Across 1000 breeding seasons, it's a different story. If that trait conveys any kind of advantage, then individuals with the trait will increase in frequency. The most important concept for us to digest here is that it takes a considerable period (years of living and breeding) for a trait to emerge as the clear winner, but eventually a population will necessarily become dominated by individuals who carry that beneficial trait. Because an adaptation reflects a response to living and breeding within a specific environment, *it also provides a window into the specific environment that gave rise to the trait.* The only way an adaptive trait makes sense is in the context of specific environmental conditions—those conditions within which the trait evolved.

Adaptations have "functions" in the sense that they help individuals survive or reproduce better than those individuals that don't have them. Only through the lens of natural selection does an adaptation (or any other pattern in the natural world) make sense. Adaptations are often unusual features associated with target species, but even if something is clearly an adaptation because it's so extreme or abnormal, the reason the trait evolved can be puzzling. For example, consider the super long nasal passage of a desert kangaroo rat. Were long nasal passages favored because they reduce water loss, or because they enhance seed detection? Whichever, it doesn't matter; what matters is the fact that the adaptation is tightly associated with living in a seed-filled, hot, dry desert environment—those are the only environmental conditions under which the feature confers an advantage and makes any sense. Polar bears have white fur for reasons that have something to do with the snowy white background that defines the only environment they occupy. The environment is clear, even if the advantages associated with a white color are not as clear. Similarly, the white-tailed ptarmigan and snowshoe hare are camouflaged in (and now require) a snowy environment. And owls like the screech-owl referenced above are remarkably camouflaged against the trees within which they roost. These coloration patterns are indicative of a long evolutionary history of each species with its respective environment. The most important corollary of this fact is that organisms do not perform as well outside the environments in which they evolved. Thus, if historically important fire regimes and environments are the holy grail that land managers are desperate to find, they ought to appreciate that those environments (and their associated fire regimes) are written all over the organisms that live in any given area; all they must do is recognize and understand the significance of the fire-related adaptations that the organisms carry.

So, which traits might be considered adaptations to specific kinds of fire? How do we even figure that out? We could look at variation in characteristics of a trait within a population to see if the typical form performs better than the atypical form, just the way natural selection does. The problem, of course, would be one of sample size and time. Natural selection evaluates millions of individuals over a period of thousands of years—we would be lucky to collect data on dozens of individuals for 1 or 2 years at best. So, that approach will probably not yield clear cut results. We could, instead, modify a trait on a subset of individuals to see if their performance (survival and reproductive success) is related to that variation. Or we could look at performance of individuals under different environmental conditions to look for environmental effects. For example, lodgepole pines with serotinous cones perform better (produce more seedlings) in severely burned environments than they do in less severely burned environments, which suggests that closed cones help when fires are severe. Even without obtaining any data on performance, identifying a correlation or an association between a trait and a certain environment is a great place to start. It we see a clear association between certain *fire characteristics* and certain *plant characteristics* that would be a sign that we probably have an interesting cause-effect relationship going on here. Unusual traits (e.g., serotinous cones that open only under extreme heat) make biological sense only when placed in the context of an appropriate environment (one with intense, large, infrequent, duff-consuming fires). Thus, the best place to start is to use simple logic.

What aspect of fire can a species become adapted to? Juli Pausas and Jon Keeley have emphasized that "…no species is adapted to a single fire characteristic; rather, species are adapted to a particular fire *regime*" [2]. To be clear, the physical characteristics of fire (its magnitude, patchiness, extent, and timing) in combination with a particular fire frequency define the environmental condition to which fire-dependent organisms are adapted. A fire-based adaptation confers an advantage to the organism only in a particular fire environment. So, how do we know it was a *fire regime* and not something else that caused the evolution of a particular trait? Just because a plant or animal occurs in a burned forest, it does not follow that the organism carries a trait specifically designed as an adaptation that enables it to do well in a burned forest. What would it take, then, to convince you that a trait is an adaptation that specifically evolved to enable an organism to survive or benefit from living under the pressure of a particular fire regime? Is a trait more likely to be considered an adaptation to fire if we simply can't think of another reason for the trait to exist (a root burl simply must be an adaptation to fire)? We can do better than that. We can muster evidence that the supposed adaptation is both *unusual* and *tightly associated with specific fire conditions*. When certain characteristics of a plant or animal occur only in association with certain fire environments, then we may be looking at traits that are truly fire dependent.

There are surprisingly few studies of the evolution of fire-adaptive traits in plants, and many plant traits have been uncritically labeled as "fire adaptations" without any rigorous analysis of how the trait functions or how the trait evolved. We need to know not only that sprouting tendency varies predictably with the fire history of a place, but that the differences reflect a genetic divergence as well. Unfortunately, that kind of rigor is almost always missing, and it is not uncommon for people to use the word *adaptation* casually and uncritically in relation to fire. For example, a Montana Fish, Wildlife and Parks employee recently stated in a Missoulian newspaper article that elk running from fire is a survival mechanism demonstrating that "…wildlife species have evolved with fire for thousands of years." It's a stretch (to say the least) that the ability to run is an evolutionary product of, or an adaptation to, living with fire! People clearly have trouble recognizing what an adaptation is. The problem is not just academic. By failing to recognize what constitutes an adaptation, one can argue (incorrectly) that the presence of some trait reflects the historical occurrence of a particular local fire regime when, in fact, the trait may not be an adaptation to that fire regime at all.

Also, much of what we see in nature are epiphenomena (correlations rather than cause-effect relationships), not adaptations. An absurd claim might be that an underground root system represents an adaptation to fire when the evolution of roots is more likely a response to the need for water uptake and plant support—any safety from heat death is merely an epiphenomenon (unusually large root crowns are another story, however). How about an animal that captures prey fleeing from actively burning fires? Is that behavior a genetically programmed adaptation to fire or simply opportunism? Does the running ability of an elk represent a "survival mechanism" that has evolved *because of living with fire for thousands of years,* or is its ability to escape fire by running simply an epiphenomenon? How about the fact that pine seeds have evolved coats allowing them to withstand intense heat? The latter seems more reasonable, but even that is a dubious claim without comparing heat tolerances of seeds that differ in coat thickness.

Most adaptation examples might seem like just-so stories, but even without a more detailed analysis, we can still uncover likely adaptations to fire by looking at the distributions of organisms in relation to fire types and regimes, or from studies of performance relative to attributes of a fire or fire regime. Some adaptations are conspicuous, and their functions are obvious; they are, therefore, enlightening. There may well be organisms adapted to (restricted to or perform well only under) a very specific kind of fire or fire regime, and that carries important management implications. Rather than providing you with an encyclopedic review of plant and animal adaptations to fire regimes, let me provide a few striking examples just to illustrate how the biology of organisms can be used to understand the kind of fire that promoted the evolution of those adaptive traits.

Consider plants first. Some adaptations enable plants to stand their ground and survive a fire event. For example, Douglas-fir, western larch, and ponderosa pine co-dominate conifer forests throughout much of the West, and they have thicker bark than any other western conifer tree species.

They are also the tree species that best survive fire. This suggests that bark thickness may be an *adaptation* to survive fire events. But what kind of fire events? It's not uncommon to hear people say that ponderosa pine is adapted (through its thick bark) to live in areas that support a frequent, understory fire regime. But did a frequent, low-severity fire regime favor the evolution of thick bark? If so, then the distribution of ponderosa pine (and the distribution of the other thick-barked species) ought to be nearly restricted to places that burn at lower-severities, and they should not perform well outside those places. It's a telling observation that Douglas-fir and western larch grow almost exclusively in mid-elevation zones in the West. Even ponderosa pine occurs throughout higher latitude, mid-elevation, mixed-conifer forests that support a mixed-severity fire regime. Therefore, the relatively thick bark in these three plant species may not have evolved from pressure to survive low-severity fires at lower elevations. Instead, thick bark may well have evolved from pressure of high-severity patches both inside and outside the low-elevation zone. Think about it. If every tree were to survive a mild fire (and most do), then relatively thick-barked individuals would do no better (and would probably even do worse) than thin-barked individuals that don't divert energy toward a structural feature that doesn't confer a distinct advantage. Uniform low-severity fire is, therefore, probably not the fire environment that favored the evolution of thick bark. For a mature tree, it would take, at the very least, locally severe (mixed-severity) fire to provide the selective pressure needed to kill many and leave only relatively thick-barked individuals to the future. So, severe fire may play a prominent role even when infrequent and/or limited in spatial extent within any given fire. It is certainly possible that differential mortality occurs primarily in the seedling stage, in which case frequent, low-severity fire might have favored individuals with thicker bark, but I'd be surprised if sizable patches of severe fire weren't the primary driver for the evolution of thick bark in all three of those tree species.

There are a couple of plant adaptations I can think of that are less ambiguously linked to a low-severity fire regime. One is the way ponderosa pines self-prune by dropping their lower branches as they grow so that, as mature trees, their canopy foliage is well off the ground. That trait makes sense because it allows individual trees to avoid having their entire canopy consumed in a relatively mild fire. Another involves the way longleaf pines grow in the southeastern United States. As seedlings, longleaf pines resemble huge bunchgrasses as their needles grow up to surround slower-growing meristematic tissue, which protects that sensitive tissue from the heat of a rapidly passing low-severity fire. Once the low-severity understory fire passes through, the seedlings then shift into high gear and grow rapidly. By the time the next fire burns a few years later, the tree is tall enough to escape damage from the flames. These growth-related adaptations undoubtedly reflect a long evolutionary history with low-severity fire and, therefore, a good indication that a low-severity fire regime was historically important in the places these plant species grow naturally.

If you continue playing this thought game, it's clear that a lot more fire-related plant adaptations are tightly associated with high-severity than with low-severity fire. For example, many chaparral plants (e.g., chamise, manzanita), western forest shrubs (e.g., maple, alder, willow, serviceberry), and eucalypts have lignotubers (burls) just below ground from which stored food reserves are mobilized after fire, allowing the plant to resprout quickly. The depth of the root crown can also account for the differential survival of species after fire and can provide a clue about the average intensity of fire that has favored the evolution of that crown depth. Most of these plants are not heavily grazed, so in these instances, the ability to resprout is undoubtedly an adaptation resulting from exposure to fire—fire severe enough to toast all a plant's aboveground vegetation. Pitch pine, some eucalyptus species, and even giant sequoias can re-sprout branches along their trunks from epicormic buds following severe fire (Fig. 11.2). Many forms of resprouting

Fig. 11.2 Eucalyptus trees have a remarkable adaptation derived from living in a severe-fire environment—the ability to resprout from epicormic buds after a severe, crown-fire event

probably represent adaptations to pressures from grazing, predation, insects, or disease, rather than adaptations to fire per se, but there's no way the epicormic resprouting high in a eucalyptus or redwood tree evolved in response to grazing pressure! There's only one kind of fire that would promote the evolution of an ability to resprout well above the ground following fire, and that is severe crown fire. The deciduous nature of western larch needles is probably also an adaptation that enables those trees to re-grow rapidly after a relatively severe fire blows through.

As with traits that decrease flammability and promote an individual's survival, traits could also evolve to increase heat levels or durations such that the plant itself ignites in flame. Self-immolation could represent an adaptation to promote fire and is most likely to occur in plants of chaparral shrublands and in eucalypt forests. How in the world could a self-destructive trait like that evolve? The composition of a plant's leaves, the volatile chemicals it harbors, or the structure of its bark (flaky) could increase the chance of death to both the parent plant and competitors around the parent plant, but that death may also allow recruitment from the same plant's special, fire-adapted seeds—seeds that require heat or smoke to germinate. Thus, a trait affecting flammability could have evolved to kill the plant, yes, but in so doing, enable individuals with the trait to recruit from seed better than individuals without the trait in environments with recurring fires, even crown fires.

Many plant species in the families Pinaceae, Cupressaceae, Casuarinaceae, Myrtaceae, and Proteaceae have cones that are basically sealed shut but they retain viable seeds that are released only after being heated, usually by relatively severe fire. Why have so many conifers evolved the combination of serotiny with cone positions being high atop the tree? Seed release following severe fire is key—just look at the recruitment of larch and pine seedlings beneath dead trees that can retain seeds in their cones for years. Some plant species have wingless seeds that are dispersed primarily after fire by birds. As I mentioned earlier, Clark's Nutcracker harvests, carries, and stashes wingless whitebark and ponderosa pine seeds in recently burned areas, providing a mechanism for the tree to colonize new severely burned sites as they become available.

Finally, there are some amazing adaptations that allow the seeds of some plant species (e.g., *Ceanothus, Rhus, Arctostaphylos, Ribes, Geranium,* and *Illiamna*) to not only survive fire, but to be able to wait for decades (for 200–300 years in the case of *Ceanothus*) before being stimulated by, and germinate in response to, the intense heat from relatively severe fire. In Glacier National Park, pinegrass produces flowers only after fire, while showy aster, dragonhead (*Dracocephalum parviflorum*), geranium, and rock harlequin (*Corydalis sempervirens*) all produce seeds stimulated to germinate after fire. It takes a severe fire to produce enough heat to allow water to penetrate the seed, and germination to ensue. Some plant species also amass a huge seed bank during the inter-fire period, resulting in a mass flowering event when fire breaks the hard seed coats and induces heat- and smoke-induced germination. Although a product of fruiting rather than seed germination, the abundance of morel mushrooms following fire is also a well-known phenomenon (especially well known by mushroom hunters). Recent research in Yosemite following the Rim Fire showed that the mushrooms occur almost exclusively in areas where the fire consumes everything along the forest floor, leaving only blackened soil, ash, or burnt needles [3]. You can thank severe fire for the morels you enjoy with an evening meal!

Now let's consider animals. Are there adaptations that clearly allow animals to thrive under certain kinds of fire conditions? Absolutely. One of the most amazing animal adaptations to living in a fiery environment involves heat-sensing organs carried by flat-headed wood-boring beetles in the genus *Melanophila* (Fig. 11.3). They have evolved heat sensors on their thorax to find still-smoldering trees that have been killed by fire, thereby rendering the trees defenseless against the beetle's attack. Eggs are deposited below the bark, after which the beetle larvae feed on the cambium of newly killed trees before they later emerge as adult

Fig. 11.3 *Melanophila* fire beetle and an electron micrograph of the heat-sensing organ it has on its thorax. (Modified from Figure 1 in Schmitz et al. 2009)

beetles. Adults are known to be stimulated by heat and/or attracted to smoke. The best fire-beetle story is one originally told by E. Gorton Linsley who noted that at University of California football games, with 20,000 or so cigarettes ablaze at any time (remember, this was the 1940s), "…a haze of tobacco smoke would hang over Memorial Stadium. *Melanophila* beetles would 'annoy patrons by alighting on the clothing or even biting' during a big game, which was more disturbing to fans than a Stanford touchdown" [4]. Linsley found that the beetles had sensory pits on their bodies and could somehow sense heat or smoke. William Evans determined subsequently that these pits were infrared detectors that allowed beetles to find burned areas by detecting heat emitted from newly damaged trees [5]. The beetles are said to be able to detect heat from objects 100 kilometers distant! Those burned trees have perfectly good wood beneath the charred bark layer but because severe fires kill them, they are unable to pitch out attacking beetles. Blackened trees are the very places where the beetles can successfully lay eggs and rear their broods. High-severity fires clearly provide that historically important burned-forest condition needed to promote the evolution of those heat sensors. The long antennae on long-horned beetles are also amazing adaptations that allow beetles to detect tiny concentrations of smoke in the air, which allows them to find still-smoldering, dead trees to attack [6].

Animal adaptations derived from living in a fire environment might include not only morphological traits like heat sensors, but also body coloration patterns and even behavioral traits. One of the most instructive behavioral traits that an animal can carry is its genetically determined tendency to settle nonrandomly by preferring some habitats and avoiding others—this is the essence of habitat selection. All animals are nonrandomly distributed with respect to vegetation community types and conditions but, in the extreme, some are relatively *restricted* in their distributions to a very narrow range of conditions. If an animal species occurs in a single environmental condition and no other then, by definition, it has evolved to *depend* on that condition. Now you might be able to appreciate the significance of my finding that the Black-backed Woodpecker is relatively restricted to and, therefore, relatively dependent upon severely and recently burned conifer forests.

But how can a plant or animal species evolve to depend on a blackened, severely burned, early postfire forest condition if such a condition occurs only rarely and briefly in any given location? There needs to be continuous and consistent evolutionary pressure from a specific environment if natural selection is going to favor organisms that carry a particular trait that allows them to do well therein. The answer becomes clear when one looks at the large-scale pattern of fire occurrence. Across the entire northern Rockies, for example, almost every year since the turn of the century has had at least one fire of 100 ha or more in size. Thus, while a severe fire may occur only once every 300 years, on average, in any one spot within that larger landscape, such fires occur very frequently across a larger area. Thus, recently burned forest patches are not rare—they are always present across a relatively large area. If a plant or animal can rapidly colonize newly burned patches of a given age within that relatively large area (as all fire-dependent species are able to do), that would keep them living (as a species) in a consistent environment. A regionally abundant burned-forest environment would produce the same kind of consistent selective pressure as that exerted upon a population living in a single place and environmental condition that is relatively unchanged over time. This same principle helps explain why few, if any, bird species have evolved to depend on blackened forests farther south than Wyoming in the Rockies and farther south than the southern Sierras in California. The more southern mixed-conifer forests receive severe fires naturally but too infrequently to create a patchwork where a burned forest is always somewhere nearby.

In the western United States, the largest number and most compelling fire-regime adaptations are linked to the higher-severity regimes that lie primarily within the mixed-conifer forests of the Sierras, eastern Cascades, and northern Rockies. Not only do different species occupy different stages of succession following severe fire disturbance in those forests, but many of those species occur under ephemeral conditions following no other form of disturbance. Animal species nearly restricted to recently and severely burned forests include the fire jewel beetle, some native bumblebee species, and the Black-backed Woodpecker. I should add that the fire morel mushroom and many plant species (e.g., Bicknell's geranium) and are also restricted in their distributions to severely burned forest conditions. Nothing reflects the naturalness and historical importance of severe fire in mixed-conifer forests more powerfully than the restricted distribution patterns carried by some species that occupy severely burned forest conditions.

The most iconic severe-fire dependent bird species worldwide is, without a doubt, the Black-backed Woodpecker (Fig. 11.4)—the centerpiece of this fire story. A long evolutionary history with naturally occurring burned-forest conditions is not only evident through its use of such forest conditions to the near exclusion of every other habitat type available throughout much of the western United States (see Fig. 4.1) but is also elegantly revealed through the jet-black color of that bird's back—that coloration pattern *screams* wildfire! Think back to my description of how natural selection works over long periods of time to favor

Fig. 11.4 The Black-backed Woodpecker is the star of this fire story. It has an uncanny ability to find and colonize severely burned forests within a year following a severe fire. More importantly, it is also nearly restricted in its distribution within the western United States to severely burned mixed-conifer forests

plumage coloration in owls that come to match their preferred resting sites. Is the cryptic black back on a woodpecker against a severely burned and blackened tree any less impressive than the ptarmigan's white plumage pattern against a snowy alpine field (Fig. 11.5)? Each of these coloration patterns reflects a long evolutionary history that involves living in a very specific environment. If the Black-backed Woodpecker story can't convince you that blackened conifer forests represent perfectly natural and historically important environmental conditions that have always

Fig. 11.5   The cryptic coloration pattern of a Black-backed Woodpecker (on left) is a clear reflection of its long evolutionary history with blackened forests, just as the white plumage of a White-tailed Ptarmigan (on right) is a clear reflection of its long evolutionary history with alpine snowfields

occurred within the bird's geographic range, then there is nothing in biology that can do so. The lesson here is that same blackened forest condition is not only natural but necessary to maintain across the broader landscape.

The only way most all the plant and animal adaptations I've described above make sense is in the context of a fire regime that includes severe fire. And remember, although severe fires burn infrequently in any one place, they burn frequently enough across a larger landscape to provide an ever present, beautifully blackened, burned-forest environment that can favor the evolution of those adaptive traits. People who argue that our western mixed-conifer forests evolved entirely in the presence of a low-severity understory fire regime are wrong. Most of our western mixed-conifer forests were born of, and are now maintained by, periodic high-severity fire [7]. I hope you see the power of natural selection when it comes to uncovering what must have been historically important fire regimes. The plants and animals in severely burned forests carry some amazing adaptations that reflect the presence of historically important severe fire in those places. They also reflect how important it is to now maintain severe fire in those same places. It seems that many of Smokey's plant and animal friends find ideal conditions in the very burned forests we have been taught to hate!

References

1. Dobzhansky T (1973) Nothing in biology makes sense except in the light of evolution. Am Biol Teach 35:125–129
2. Pausas JG, Keeley JE (2009) A burning story: the role of fire in the history of life. Bioscience 59:593–601
3. Larson AJ, Cansler CA, Cowdery SG, Hiebert S, Furniss TJ, Swanson ME, Lutz JA (2016) Post-fire morel (Morchella)

mushroom abundance, spatial structure, and harvest sustainability. For Ecol Manag 377:16–25
4. Linsley EG (1943) Attraction of *Melanophila* beetles by fire and smoke. J Econ Entomol 36:341–342
5. Evans WG (1966) Perception of infra-red radiation from forest fires by *Melanophila acuminata* DeGeer (Coleoptera: Buprestidae). Ecology 47:1061–1065
6. Schütz S, Weissbecker B, Hummel HB, Apel K-H, Schmitz H, Bleckmann H (1999) Insect antenna as a smoke detector. Nature 398:298–299
7. Alt M, McWethy DB, Whitlock C (2025) Postglacial fire and vegetation histories of a mid-elevation mixed-conifer forest in the Gallatin Range, MT, USA. Quat Res:1–15

12

Why the Big Secret?

Adaptations representing evolutionary responses to fire are now carried by organisms that, in most cases, have come to depend on the kind of fire or fire regime against which the adaptation evolved. Thus, having an adaptation to fire (e.g., a tree that can shed needles, withstand high heat, or flower profusely, or a bird that can home in on a newly burned forest where beetle larvae are abundant) not only allows a plant or animal to survive or take advantage of fire's creation, but also reflects the fact that the organism now *needs* a certain kind of fire to persist. Severe fire is the *only* disturbance agent capable of sustaining much of the diversity of plants and animals in western conifer forests…that's the most important take-home message here.

The more time I spend in burned forests, the more the wonder of it all amazes me. Even so, after years of field work and publishing, I came to a growing realization that the benefit, exquisite nature, and ecological necessity of severely burned forests remains hidden from public view. Most

R. L. Hutto, *A Beautifully Burned Forest*,
https://doi.org/10.1007/978-3-032-03180-8_12

people, including most fire scientists and land managers, are still unaware of how important severely burned forest conditions are for so many species. Why is that? Why has this amazing story remained for so long under a veil of secrecy? There are several possible reasons, including the inadequacy of my own story telling.

In my own case, it's certainly not from the lack of trying to tell this amazing story. I chose Conservation Biology, a very applied science journal, to publish my initial research results, which showed just how biologically unique severely burned forests are. In the three decades since that 1995 publication, I have retold the basic story in one form or another through outlets that included "Birdwatch" (a PBS television show that I hosted), local television and radio news, newspaper op-eds, professional meetings and symposia, invited talks, field trips, dozens more peer-reviewed papers, and web-based outlets, including a Facebook page [1], a fire ecology page [2], and a YouTube channel [3]. Other writers have also written about the fire story in book chapters, magazines, children's books, nature trail guides, and interpretive signs. I even went through a years-long process to get the Montana Board of Regents' approval in 2004 to establish the Avian Science Center on the University of Montana campus, a center whose mission was (and still is) to "…disseminate research and monitoring results in a manner that promotes conservation of natural resources through informed land management and land stewardship practices." Still, after all that, evidence from every angle seems to suggest that few people have ever heard about the biological uniqueness, naturalness, and necessity of severely burned forest conditions.

Even though I have tried to spread the fire story broadly, one possibility is that I (and other writers) fail to communicate science effectively. A couple of decades ago, other scientists who conduct heavily applied research were experiencing the same frustration with their inability to transfer scientific information effectively, as evidenced by an explosion of interest in science communication at that time. After about 2000, I began to see more and more opinion pieces in journals, more books on science communication, special sessions at professional meetings and the like. Prominent scientists were warning that it is no longer enough just to do science, and that "…knowledge must be conveyed in a way that allows policy makers and the public to translate the science into action" [4]. Thus, I began a concerted effort to do a better job of "science communication." I even participated in a 2014 wildfire science communication workshop in Seattle organized by COMPASS, an independent organization devoted to helping scientists communicate effectively. The workshop was funded by the Wilburforce Foundation and was designed to bring a bunch of fire scientists together to discuss ways to better communicate our findings to the press and others.

The organizers of the Seattle workshop were careful to invite scientists from across the spectrum, from those who believed "our forests are out of whack" to those who felt "our forests are fine," so there was also an overtone of "Who's right?" throughout the workshop. We came to appreciate that who's right depends heavily on the system in question. I may have had a great story, but it didn't apply to the dry forest types studied by most of the scientists who attended. The dry conifer forests are where severe fires burn much more frequently and much less severely, so they do not support populations of most of the fire-dependent plant and species that became the focus of my fire story. The kind of postfire magic I had experienced was restricted to the mid-elevation, higher latitude mixed-conifer forests, as reflected in the distribution of Black-backed Woodpeckers.

In a nutshell, all of us scientists at that workshop were "right" in our own worlds; the fire story is nuanced, and one size does not fit all. That could be the primary reason why the severe-fire story never gained much traction—it simply doesn't apply to most western forests. But that explanation falls short. Although the magic associated with a severely burned forest does not apply everywhere, it still applies to *most* western conifer forests, especially mixed-conifer forest throughout the Sierras, inland Pacific Northwest, and northern Rockies. At the end of the day, the story is not all that restricted geographically, so a lack of relevance to most western mixed-conifer forests is not the problem. Ineffective science communication is not the problem either. My ability to communicate may have improved after that workshop but making my story simple enough that people can understand it was not then, and is not now, the problem. The story is not at all complex; it is sufficiently clear that anyone can understand it. The most concise telling of the story is still perhaps my own 10-min video, which was filmed some 30 years ago, when I hosted the "Birdwatch" television series (that clip is now posted on a Woodstock Community Television web page) [5]. If neither applicability nor communication is the problem, then why have people not heard about, and then come to celebrate, the benefits of severe fire? I can think of five additional possibilities, and they all play at least a small role.

One reason is that fire is inherently dangerous, and danger seems to dampen or decrease the volume on anything else that might be true about fire. When something can kill you, it seems responsible to focus on demonizing that entity and not to say fire is a good thing. But there's something else going on here. We don't demonize the sun, even though its rays can kill us! That is because everyone understands how the benefits of sunlight far outweigh the potential costs, which we minimize by wearing clothes and using sunscreen. Unlike the sun, most people's perception of fire is different because the benefits are hidden from public view and the costs are overblown by most fire scientists and the popular media. The most thorough assessment of fire risk is probably that of Jack Cohen, a retired fire scientist at the USFS Fire Science Laboratory in Missoula [6]. His telling research results are clear, if shocking, to folks both in and out of the fire community—*fire danger to human communities is nearly entirely the product of conditions of a home and its immediate surroundings, not to fuels miles away, or even vegetation conditions as close as 50 meters away.* This is not to say fires carry no risk; it's just that risk is surprisingly small if communities have taken the effort to become "Firesafe." If people could adjust the level of risk downward to something close to reality, then perhaps stories about the positive side of severe fire would emerge above the noise level surrounding safety.

A second reason the severe-fire story might not have gained traction is that, even though there are dozens upon dozens of published fire effects studies, research designs are often inadequate to the task of revealing the story. If most field studies are not designed in a way that would reveal the biological uniqueness associated with severe fire, we won't see the story repeated across every new scientific publication on fire effects. One must sample across a large range of vegetation types and conditions to show that a species is restricted to a small subset of burned-forest conditions. That takes a special kind of long-term and geographically extensive research design involving not just one study site, but hundreds of sample locations. When I began my own research in 1988, the studies of birds in burned forests be counted on one hand and none involved geographically extensive, multiple-habitat surveys. The absence of habitat-relationships studies persists to this day. One would think that, given the emergence of coordinated landbird monitoring programs, interesting patterns of association between bird species and specific environmental conditions are now more likely to be exposed. Unfortunately, even today, most bird monitoring programs are not designed to uncover habitat relationships; they are designed to uncover population trends. Interesting habitat relationships can be exposed only if the bird survey data are tightly linked to specific vegetation types and conditions. Without habitat relationships, patterns are hidden from scientists themselves and from the public as well.

Another research design problem is that the methods most scientists use to "test" for fire effects on plants and animals are often inadequate to the task. For example, very few of the early publications on "fire effects" describe the nature of the fire, even though effects vary markedly with time since fire and fire severity. This is not unlike ignoring the dose of herbicide in a study of herbicide effects. Unlike the situation with herbicide studies, which would never be published without information on the kind or dose of herbicide used, fire studies are routinely published without any reference to the kind or dose of fire studied. The most ecologically significant fire effects may appear in response to only a narrow range of fire types—a plant or animal trait could be adaptive in one type of fire and maladaptive in another, so we ought to know the kind of fire a person studied. We underscored this important concept in one of our own papers by emphasizing that biologically significant "fire effects" cannot be exposed by lumping all one's data from burned forests into a single pot to compare with unburned forest data. Results are *context dependent*—they depend on pre-fire vegetation type and condition, fire severity, and time since fire (among other things) [7]. In other words, a result that emerges from a low-severity burn might not hold true when the severity is greater. The *kind* of fire matters, and research results are sensitive to that fact.

A final research design problem I'll mention is that scientists often use some measure of "diversity" as a response variable in their studies of fire effects. Unfortunately, diversity indices always hide more than they reveal about individual species responses because diversity can be unchanged while the populations of any number of key species can change dramatically. This loss of information associated with the use of multi-species summary statistics is almost a crime, but the use of diversity as a fire-effects metric (indeed, as a general conservation metric) persists to this day. Telling us that diversity did not change tells us nothing about whether there were significant fire effects hidden amid that lack of change in diversity. If one wants to look for biological effects of fire (or the effects of any environmental parameter), one must understand it is *individual species* that reveal those effects, and those effects are almost never captured by employing some multi-species metric as a response variable.

Even when an interesting severe-fire story does emerge from well-designed research and analysis, those who publish results that carry negative management implications will face yet another information-dissemination problem. This problem (and the third reason it's hard for the fire story to get the attention it deserves) is that there will be backlash from some scientists if research results run counter to the dogma that western conifer forests are out of whack and need thinning, prescribed burning, and planting. Many fire scientists are strongly invested in the idea that forests are out of whack and that forest restoration through thinning and prescribed burning is desperately needed. Consequently, there is also no shortage of what are basically opinion pieces in professional journals claiming there is a significant risk associated with severe fire which can be alleviated only through more thinning and burning [8]. It is no coincidence that the same scientists are funded, in large part, by the fire-industrial complex (the publication I just cited in the last sentence was, for example, paid for by CalFire). One of the most fascinating aspects of reporting research results that run counter to the idea that our forests are out of whack is that you will find yourself swimming in a river with very strong currents, and getting your hands high enough in the air to attain visibility is going to be a chore. Disagreements are not uncommon in the scientific literature, but the tone of attacks and the movement of attacks into newspapers and other media outlets is noteworthy. Emily Guerin is one of the first to expose this deep riff in the fire research community in a 2012 article she wrote for High Country News. A follow-up conversation with Guerin about the two camps of scientists was also recorded in a podcast [9]. The podcast is a brilliant piece of work that deserves everyone's attention because most people don't realize how contentious things have become in the world of fire research. The intolerance of "party-line" scientists to any viewpoint that runs counter to the "forests are out of whack" dogma (even results based on elegant scientific research) is eye-opening.

A fourth reason the fire story remains hidden is that fire and wildlife professionals stand in the way by restricting their discussion of wildlife benefits associated with high-severity fire to benefits that are nice but not at all necessary. It's not compelling when a forest manager or a forest biologist notes that elk, American Robins, and House Wrens derive a benefit from fire. Those species use burned forests opportunistically; they occur predominantly in forests that have not burned and would do just fine without fire. Thus, their choice of wildlife species conflates the "benefit" of fire with the "necessity" of fire. By highlighting the fact that deer move into burned areas, they unknowingly interfere with the public coming to understand and appreciate that there are plants and animals that not only use but *require* those severely burned forest conditions. Noting that a species like the Black-backed Woodpecker is nearly restricted to burned-forest conditions *is* compelling. Many plant and animal species don't simply benefit from severe fire—they need it. That fact ought to garner the attention of land managers because they are charged with maintaining the ecological integrity of the systems they oversee. Even professionals could do a better job of placing the emphasis where it ought to be concerning the benefits of fire.

The four reasons outlined above certainly play roles in keeping the fire story hidden, but the combined effect of all those possible explanations still pales in comparison with what I believe is the single most important (fifth) reason this story remains a secret—land management agencies (and even conservation organizations) have not yet come to embrace severe fire because doing so would run counter to what they see as their mission. My fire story is an easy-to-understand but politically unpopular story. And it's not just that many land management agencies and scientists fail to embrace the benefits of severe fire. It's worse than that. Their public messaging is specifically designed to demonize severe fire. Just look at Smokey Bear's message or any other public message surrounding wildfire. One of the most ecologically uninformed attempts at public messaging by the USFS appeared in on a billboard in my own town upon which the text "Unfortunately, *they* can't run for their lives" was superimposed on a picture of blackened, standing-dead trees following a severe fire (Fig. 12.1). That was the message, even though those trees *depend* on severe fire for their

Fig. 12.1 Typical USFS messaging regarding severe fire within fire-dependent forests. What do signs like these make you want—a green-tree forest or a blackened, naturally restored wonderland? I was particularly taken by the ecologically uninformed messaging in the billboard suggesting that fire-dependent tree species can't run from fire—the very thing they need for reproductive success!

reproductive success! NGOs and governmental agencies swamp us with information about how bad severe fire is through their use of websites, webinars, and pamphlets with titles like "Science You Can Use." Public land management agencies (and at least one prominent conservation organization) are marching lockstep with a large and vocal group of "forests are out of whack" scientists. They have, so far, failed to acknowledge results from key studies that signal a strong need to inform the public about the ecological importance of severely burned forest conditions. Even worse, these organizations purposely paint severely burned forest conditions as unnaturally bad outcomes of a mismanaged forest. The reason for doing so seems obvious. The "forests are out of whack" story supports ulterior management motives to "fix" perceived forest health problems by using the huge industry surrounding fire and timber management. Management practices are strongly tied to legislation passed down from people who are not well versed in forest ecology. Many legislators believe (incorrectly) that severe wildfires are incompatible with the maintenance of a healthy fire/timber industry, so they won't allow a story about the benefits of severe fire to gain traction. Instead, we are inundated with message after message telling us that a severe fire leads to a bad outcome compared to what a mild, understory fire brings (Fig. 12.2). I was taken aback when I even found an illustration of this kind of misleading "good fire/bad fire" message in a recent scientific publication by reputable fire scientists [10].

Over the years, I have taken hundreds of people (including district rangers and forest supervisors) into a severely burned forest to witness the biology therein. Every one of them grew up bathed in the message that severe forest fires are bad, and every one of them had been led to believe through cherry-picked research and widespread public messaging that only low-severity understory fire is good for a forest. Nevertheless, I have found that a person's perception of severely burned forests can change a full 180° during a 30-min walk through a blackened landscape. This kind of "fire walk" is the only form of science communication that has worked for me, but that form of communication does not reach the masses. We're left with the fact that most

Fig. 12.2　This published illustration (from Kaufmann et al. [11]) is clearly designed to suggest that there are only two options for land managers—either thin and prescribed burn (as in the left side of the photo) or suffer the consequences of severe fire (the right side of the photo)

people have never been shown the beauty associated with, and the dependence of so many species on, severely burned forests. I sometimes think that the story I'm telling here is nature's best kept secret. Unfortunately, when the value of severely burned forests is hidden even from those who manage our public lands, then current management practices may even interfere with the creation and maintenance of severely burned forest conditions. I hope this book can help turn the tide because once one understands the fire story and the ecological necessity of severe fire, it becomes much easier to recognize what I believe are several key threats to maintaining an intact disturbance-dependent forest system. I outline those threats in the next chapter.

References

1. https://www.facebook.com/FireEcologyLab. Accessed 1 July 2025
2. https://fire-ecology.com/. Accessed 1 July 2025
3. https://www.youtube.com/@fireecology. Accessed 1 July 2025
4. Palmer MA, Bernhardt ES, Chornesky EA, Collins SL, Dobson AP, Duke CS, Gold BD, Jacobson RB, Kingsland SE, Kranz RH, Mappin MJ, Martinez ML, Micheli F, Morse JL, Pace ML, Pascual M, Palumbi SS, Reichman OJ, Townsend AR, Turner MG (2005) Ecological science and sustainability for the 21st century. Front Ecol Environ 3:4–11
5. https://www.youtube.com/watch?v=0PTJ9wbj9_0. Accessed 1 July 2025
6. https://www.youtube.com/watch?v=ubybKQ6ligY&t=201s. Accessed 1 July 2025
7. Smucker KM, Hutto RL, Steele BM (2005) Changes in bird abundance after wildfire: importance of fire severity and time since fire. Ecol Appl 15:1535–1549
8. Williams JN, Quinn-Davidson L, Safford HD, Grupenhoff A, Middleton BR, Restaino J, Smith E, Adlam C, Rivera-

Huerta H (2024) Overcoming obstacles to prescribed fire in the North American Mediterranean climate zone. Front Ecol Environ. https://doi.org/10.1002/fee.2687

9. https://www.hcn.org/blogs/articles/west-of-100-fire-brimstone. Accessed 1 July 2025

10. Stephens SL, Westerling AL, Hurteau MD, Peery MZ, Schultz CA, Thompson S (2020) Fire and climate change: conserving seasonally dry forests is still possible. Front Ecol Environ 18:354–360

11. Kaufmann MR, Shlisky A, Marchand P (2005) Good fire, bad fire: how to think about forest land management and ecological processes. Pages 1–12. U.S. Department of Agriculture, Forest Service, Rocky Mountain Research Station, Fort Collins, CO

13

The Hidden Fire Story Conceals Management Threats

Public land management agencies are required by laws like the National Forest Management Act and the National Environmental Policy Act to maintain the ecological integrity of the lands they manage. That would include maintaining severe fire and the resulting conditions as products of the natural disturbance process. Given these laws, it seems fair to ask whether land managers are currently doing everything they can to allow the *creation* and *maintenance* of those ecologically special burned forest conditions that follow severe fire. It doesn't seem like they are. Instead, they seem to be obsessed with mitigating (lessening) severe fire effects through pre-fire timber harvesting and prescribed burning. When those efforts fail (and they nearly always do), then we spend billions of additional taxpayer dollars on attempts to suppress the fires that do manage to burn severely. And after they fail to suppress wind-driven, severe fires (as is most often the case), management efforts immediately shift toward salvage logging the blackened forest

139

R. L. Hutto, *A Beautifully Burned Forest*,
https://doi.org/10.1007/978-3-032-03180-8_13

that just got created. Salvage logging is then followed by tree planting and sometimes herbicide spraying to speed "recovery" of the forest after severe fire. To me, this does not seem to reflect enlightened ecosystem management.

It sometimes seems like land managers don't even want to hear about potential ecological costs associated with a proposed management plan. They try their best to thwart any public criticism of pre- and postfire logging plans by inserting ill-defined, mother-love terms into their written plans. Just read any land management plan related to a timber harvest. The plan will include words such as "resilient" and "forest health" and "variety," and will also emphasize "community safety" goals that are largely unrelated to forest condition. It is hard to argue against goals based on ill-defined terms that are never measured. Land managers can do better than this. They can still harvest timber and fight fires but do so in ways that largely eliminate the most significant ecological costs to the disturbance-dependent conifer forestlands that grace the West. There is a way toward a better and more honest use of our natural resources, but the path first requires that *ecology* be positioned squarely in the driver's seat.

Ecology is not currently in the driver's seat, however, so ecologically damaging management practices are not only common now but are on track to become even more widely employed across forest lands in the face of recent federal legislation designed to address our nation's ailing infrastructure and in the face of a new administration that will redouble that effort. Let me explain how current management emphases before fire (i.e., pre-fire timber harvesting and prescribed burning), during fire (i.e., fire suppression), and after fire (i.e., postfire salvage logging and tree planting) act against the creation and maintenance of ecologically special, severely burned forest conditions. There are stated justifications for these management activities, of course, but the most common justifications are weak at best, as I will explain. In my final chapter, I will offer thoughts on how we could mitigate the ecological threats associated with each of these management categories and still receive all the goods and services that our national forests provide.

Pre-fire Thinning and Prescribed Burning

In conifer forests, pre-fire thinning and prescribed burning are management practices designed to reduce fuel loads, and they have gained a huge head of steam recently. In what way is green-tree thinning and prescribed burning a threat to the creation and maintenance of suitable burned-forest conditions? In a way you might not have thought about. *A harvested mixed-conifer forest that subsequently burns provides nowhere near the benefit that a burned, unharvested mature forest does to the most fire-dependent species* [1]. That's right. The plant and animal species that most depend on severely burned, standing-dead forests are nowhere to be found when restoration-thinned, mixed-conifer forests finally do burn in a natural wildfire. My guess is that most forest managers are entirely unaware of this delayed ecological cost associated with green-tree thinning and prescribed burning in mixed-conifer forests. The idea that you must have densely stocked, mature-forest conditions *before* fire to create that magical blackened condition that so many species have evolved to depend on *after* fire certainly took me by surprise. And I am someone who has worked on fire ecology for decades! Think a little more about this. A thinned forest that subsequently burns severely does not meet the needs of fire-adapted species because the legacy of mature-forest structure (e.g., tree sizes and densities) existing prior to fire affects the suitability to fire specialists after fire disturbance. Every published study of birds in burned conifer forests has shown that the most fire-dependent woodpeckers need high densities of blackened trees (hundreds per ha) larger than about 30 cm in diameter. You can't burn a recently cut forest (or even a young forest) and get anywhere near the positive response you get from the most fire-dependent species soon after an uncut forest burns severely. Forest thinning not only diminishes the suitability of those forests to fire-dependent species if those forests still manage to burn within a decade or two after thinning. Thinning also prevents those patches of land from becoming suitable to fire-dependent species for many, many decades (until those patches mature, fill in with large trees, and then burn once again).

A mixed-conifer forest that experiences a wildfire following forest thinning and prescribed burning may still look green, as celebrated by the Nature Conservancy (Fig. 13.1), but that forest will not be in the blackened condition it ought to be in or have the density of large trees it ought to have following fire. As long ago as 1938, Hoyes Lloyd noted that "Where fire follows logging, …there can be no claim of 'natural' conditions and even the temporarily beneficial conditions for wildlife may be lost" [2]. Mesic mixed-conifer forests (and probably most dry mixed-conifer forests as well) require mixed- to high-severity fire, not low-severity understory fire, and they require blackened, densely stocked, mature trees to satisfy the needs of the most extremely fire-dependent plant and animal species; it's as simple as that. It is probably more accurate to think of thinning and burning as a form of "restoration prevention" for most

Fig. 13.1 The forest condition in the middle (following thinning and prescribed burning) may look great to the ecologically naïve, but the ecological condition of a burned forest that receives that kind of treatment will not be in the condition it ought to be in following fire. There is only one place in this picture where all the post-fire specialist species can be expected to occur—in the blackened place on the right that received no treatment! (Photo Steve Rondeau)

conifer forest types because most of them need severe fire to undergo natural succession.

Even if a land manager or fire scientist were to acknowledge this largely hidden ecological cost associated with thinning and prescribed burning in forests that support a mixed- to high-severity fire regime, that same person would also likely argue that the ecological cost is worth paying when it comes to restoring forest conditions and gaining an increase in home safety. We must thin to be safe. The justification seems solid on the surface, but there are two problems with the safety justification: (1) even if forest structure were out of whack, fuel reduction will not moderate fire behavior to any significant extent during hot, dry, windy periods when fires burn across most of the land that's going to burn in any given year; and (2) fuel loads have little to do with fire risk to homes.

Concerning the fuel reduction problem, Scott Stephens, a fire scientist at Berkeley, claims that there's at least 99% certainty that treated areas moderate fire behavior "…at least 99%. I'll be honest with you, it's that strong; it's that strong" he said to a reporter recently [3]. The argument about whether forest thinning and prescribed burning reduce future fire risk in any meaningful way has gotten nastier and nastier [4]. The shouting match is entirely misdirected, however, because it focuses on whether treatments affect fire behavior most of the time rather than on whether fire behavior is affected when fires burn most of the area. If vegetation condition is the primary factor determining how much area and how severely a fire is going to burn, then we might be justified in cutting to reduce fire danger; if weather is the determining factor, then unless we address climate change, we can't do much about it. We know that both fuel and weather conditions play roles in how severely a fire will burn, but fuels play the stronger role only under moderate weather conditions (which may very well be 99% of the time, as Stephens claims). Unfortunately, fuels play a trivial role under extreme weather conditions, which is when fires tend to burn severely across 98% of forested lands that do burn in the West, thinned or not [5]. Just look at photos of thinned forests that subsequently burn severely; even seed-tree cuts don't moderate fire severity when the winds blow (Fig. 13.2)!

Extreme weather conditions (e.g., high winds and low humidity during severe drought) rather than quantity of woody fuels exert the greatest influence over fire occurrence and fire behavior, especially in mixed-conifer forests, where cutting cannot moderate fire occurrence or behavior [6]. All told, there's little evidence that excessive fuels are causing larger burn areas. So, the question is not whether treatments might work; I am certainly one of those scientists who agrees that treatments work when the weather is not extreme. The question is whether thinning moderates fire behavior to an appreciable extent when conditions are extreme, which is when fires burn most of the land that is going to burn in any given year. More importantly, the argument about whether thinning affects fire behavior is also misdirected because it does not circumvent the first problem of whether we should even be *trying* to moderate fire behavior in forests that need severe fire. Mixed-conifer forests have evolved with severe fire, and numerous plant and animal species now depend on those fires to create the burned-forest conditions they need.

Concerning the second problem with the safety argument (fire risk to homes), it's true that fire needs fuel, but it's also true that fuel loads have little to do with safety risk. As I outlined in the last chapter, human and home safety is largely a product of the home's condition and of conditions immediately surrounding the home, not a product of forest

Fig. 13.2   As shown here, even a forest that is heavily thinned prior to fire is not immune to crown fire. This is because it is weather conditions, not fuels, that dictate fire behavior most of the time in mixed-conifer forests throughout the West. (Photo George Wuerthner)

condition far beyond the Wildland-Urban Interface. Fuel-reduction and suppression efforts will do little to dampen fire spread or to eliminate the creation of blowing embers during a windy, severe-fire year if there's any fuel at all (and there will always be)—it is blowing embers (and not walls of flame) that ignite homes. For this reason, entire neighborhoods can burn to the ground while, at the same time, nearby fuels remain unburned. Creating fire-safe communities is the only way to reduce fire risk. The 2022 film, "Elemental: Reimagining Wildfire" captures this fact in a compelling fashion [7].

Safety issues aside, we shouldn't even be trying to "restore" forests that are well within their natural range of natural variation. Western mixed-conifer forests are characterized by mixed- to high-severity fire regimes that need to be maintained. The point of this book is to show how severely burned conditions are a perfectly natural and important successional stage in our western mixed-conifer forests. It is only in the dry, low-elevation forest types that restoration activities carry at least some ecological justification. Outside those dry forest zones (and that includes most areas where pre-fire thinning and burning occur), these activities are neither ecologically justified nor effective. When you thin and burn to "restore" forest conditions in most mixed-conifer forest types (places outside dry forest types for which that practice was designed), the artificiality of such treatments becomes apparent. These kinds of restoration treatments applied to mixed-conifer forest zones don't make sense. And it's not just the open-grown appearance that looks out of place; the plant and animal communities following restoration treatments do not come close to resembling those common to naturally open, dry-forest conditions either. We looked, for example, at the bird species before and after a restoration thinning treatment within a mesic mixed-conifer forest in western Montana (in an area well outside the Wildland-Urban Interface) and found that the same species occupied the sites before and after the treatment [8]. That seems to indicate that the project justification (a need for restoration) was inappropriate because we did not see colonization of the artificially open forest by bird species typical of naturally open, dry forests after treatment. The treated area was never a proper dry forest in need of restoration in the first place. Thus, when applied outside the naturally open parts of dry-forest types, there is no evidence that thinning and burning offer any kind of "restoration" that then allows an understory fire to act "…as an ecological process in fire-adapted ecosystems," as claimed by several authors in their synthesis of research on the effectiveness of such treatments [9].

Now let's address the ecological threats associated with a practice that frequently goes hand in hand with pre-fire thinning—prescribed burning. The prescribed burning part of a thin-and-burn restoration treatment is fraught with ecological costs when applied in mixed-conifer forests. It doesn't take rocket science to know that any kind of fire applied at the wrong intensity, in the wrong season, and at the wrong frequency is misguided at best and a crime at worst. The season of fire can affect a plant's resprouting ability; if fire burns just after a plant's growth period, there's no energy left to resprout. If fire occurs during spring, seeds may germinate after reaching stimulating temperature, but then dry and die during summer drought, or be eaten by masses of young seed predators that are typically produced in spring as well. As another example, the response of wood-boring beetles to fire depends on when fires occur, and fires outside the normal burning period are also outside the optimal period for driving woodborer population growth

[10]. Just remember, those beetles are among the most fire-dependent species in the West.

There is also potential ecological risk associated with burn frequency—there can be too much of a good thing. Remember, the most fire-dependent plants and animals evolved to depend not only on fire severity, but on fire frequency as well. Fire can be too frequent and not allow plants to grow enough to mature and set seed, thereby slowly depleting seed bank until local extinction takes place. This "immaturity risk" is the risk that fire frequency will be too high. Consider an obligate seeder like *Ceanothus*, which needs time to mature before producing seeds after a fire; the seeding strategy also allows it to persist in places with long inter-fire intervals as a seed rather than as an adult [11]. Or consider the conifer species in Glacier National Park that need time to develop sufficient seed stores between subsequent fires [12]. Black spruce is also having trouble regenerating after what has become too-frequent fire in the boreal forests of Canada. The specific conifer forest problems cited above are not related to prescribed burning, but immaturity risk is still a potential cost that we may be overlooking in prescribed burning programs. The bottom line is that the absence of ecological understanding concerning fire frequency and severity is glaringly apparent in prescribed fire programs.

As an attempt to add weight to the need for prescribed burning, the restoration justification has morphed recently. We now need more prescribed burning to return to conditions created by past indigenous burning practices [13], even though indigenous burning was localized, frequently conducted during ecologically inappropriate times of year, and never widespread across most western forests [14]. The most appalling thing related to the indigenous burning argument is that anyone who disagrees with the idea that we ought to implement broadly now what indigenous people did locally in the past is a culturally insensitive racist! The evolutionary history of plants and animals is clear—severe fires were the norm in conifer forests throughout most forested land in the West, and severe fire is the only kind of fire that will sustain native plants and animals. Localized burning to promote the growth of plants used for baskets or to attract or drive game by Native Americans was appropriate, but it was also utilitarian forest management; it was never widespread ecosystem management designed to maintain the ecological integrity of western forest systems. Prescribed burning outside historically important burning periods and forest types is not only ecologically inappropriate in terms of placement, intensity, and timing, but also inappropriate in terms of the way it extends unhealthy air quality by months every year.

Fire Suppression

Fire suppression outside the Wildland-Urban Interface is the second management practice that does not serve to create or maintain severely burned forest conditions. Most people believe if we detect and put fires out fast, the forest will be saved from the destruction associated with a natural disturbance event, and our homes and communities will be safe. Not only is such a goal unattainable, but it is undesirable because living organisms need fire—all kinds of fire. We may as well be trying to stop the sun and the rain or any other natural physical process that the forest needs! Fire suppression is misguided not only because it represents an attempt to thwart an ecologically critical natural process, but also because it is largely unnecessary if people are educated, if communities are "Firewise," and if homes are hardened [15]. With adequate preparation, wildland fires pose little threat to people and structures. Support for structural firefighters well within the Wildland-Urban Interface is one thing but attempting to suppress fire in places far removed from human habitation where they pose little risk to structures is costly, ineffective, dangerous, and ecologically uninformed [16].

Beside the basic fact that firefighting in the outback acts to thwart forest restoration through a natural disturbance process, the tools firefighters use only compound those negative ecological effects. Vegetation mastication, bulldozing, and dropping fire retardant are among the most ecologically costly firefighting tools. I realize how the use of aircraft to spread fire retardant is wildly popular when fires burn, but the safety/ecological cost ratio may not justify this kind of firefighting activity, especially in places far removed from urban areas. As I write, there is a lawsuit in Montana that revolves around the use of fire retardant, accusing the USFS of violating the Clean Water Act. We now drop more than 50 million gallons of ammonium phosphate-based fire retardant across federal, state, and private lands every year, and the chemicals are toxic to native fish [17].

Consider the justifications used to fight fire. One justification for fighting forest fires is that fires are bad for forests, so we need to nip them in the bud or else our forests will burn. If a little disturbance ecology seeps into the conversation, land managers may come to acknowledge the benefits of fire, which weakens the "bad for our forests" justification. Absent that justification, they will then move on to claim that, without full suppression, small fires will grow to be large fires, putting human properties and lives at risk. Perhaps the most telling fact about the latter justification is that, as embarrassing as it may be, suppression efforts do not keep tiny fires from becoming huge conflagrations.

Most people seem to believe that fires gain speed much like steam locomotives do—if we don't attack fires when they're small, they're destined to become large and destructive to human communities. But that's not true. The best anecdotal evidence emerges from Yellowstone National Park's wildland fire policy, which existed from 1972 until the fires of 1988. In that 16-year period, fires were not suppressed, and 235 fires burned a total of about 15,000 ha. Only 15 of those fires were larger than 40 ha. The larger fires were still relatively small, and all fires extinguished on their own. Less anecdotally, several fire scientists published an eye-opening study where they compared the distribution of fire sizes in one part of Ontario, Canada, which was historically fire suppressed, with another part of the same province where fires were historically left to burn [18]. The graphic they produced was astounding. In the part of the province where fires were suppressed, there was a huge increase in the proportion of tiny fires (<40 ha) because, as expected, firefighting kept most fires from growing beyond a few hectares. Here's where the results get interesting. Did the increase in tiny fires result in a large decrease in huge fires? No! The biggest relative decrease in numbers of fires came from the next-to-the-smallest fire size category, not from the truly large fire size categories. The proportion of fires belonging to the largest fire size classes (fire sizes beyond 40 ha) were nearly identical in controlled and uncontrolled areas. That is, suppression efforts merely kept tiny fires from becoming slightly less tiny; they didn't keep tiny fires from becoming truly large fires. Even though we successfully suppress 98% of all fires, those are not destined to become huge fires— they are destined to become only slightly larger fires.

These data are embarrassing to those who work in the fire-industrial complex, but facts are facts—fire suppression activity does not accomplish what we all think it does. The justification that we need to fight fires or else they will grow to be large fires is simply unsupported. Fires that are destined to become large occur under weather conditions that are not conducive to firefighting anyway—when winds are severe, conditions are simply too dangerous. With huge federal budgets and an entire industry built up around fire suppression activity, this will be anything but easy but it's high time for a change in both messaging and activity surrounding fire suppression.

Postfire Salvage Logging

The third fire-related management practice that interferes with the creation and maintenance of severely burned forest conditions is postfire salvage logging. You might ask, What's the problem with "salvaging" a little timber from a forest that has been "destroyed?" As I outlined in detail earlier, most conifer forests are not destroyed by severe fire. Many plant and animal species occur rarely, if at all, outside severely burned forests harboring an abundance of mature-sized, standing-dead trees; you can't cut them down and retain anything close to an intact disturbed-forest system. Legislation designed to expedite salvage logging and planting in the name of "restoration" (nowadays with categorical exclusions from NEPA analysis) only enables a kind of land management that has uniformly negative ecological consequences. The overwhelmingly negative effects of postfire salvage logging on the most fire-dependent organisms are, to this day, among the strongest and most consistent scientific results ever published on any wildlife management issue.

Over a quarter-century ago, I wrote that "…salvage cutting may reduce the suitability of burned-forest habitat for birds by removing the most important element—standing, fire-killed trees—needed for feeding, nesting, or both by the majority of bird species that use burned forests" [19]. Standing-dead trees are critical—73% of all cavity nests were in broken-top conifers or aspen, even though they make up only 8% of the trees. And it's not just nesting requirements that need to be met, woodpeckers need hundreds of trees per ha to find burned forests suitable. The biology of the most fire-dependent bird species suggests that even a cursory attempt to meet their snag needs would preclude postfire salvage logging in severely burned conifer forests. Two decades ago, I also wrote, "I am hard pressed to find any other example in wildlife biology where the effect of a particular land-use activity is as close to 100% negative as the typical postfire salvage-logging operation tends to be" [20]. This is not hard to understand—if you remove the trees, the most fire-dependent species will disappear (Fig. 13.3).

You must not misunderstand the big point here. I am not concerned that some woodpecker species needs severely burned forests to thrive; it's that an entire ecosystem needs a severe-fire restoration treatment every so often—the woodpecker is merely an *indicator* of whether the much larger ecosystem is intact. If the indicator species is present and doing well, then the larger ecological system that the indicator needs is probably also intact. It is, after all, the intact functioning ecosystem (ecological integrity) that land management agencies are charged to maintain, not just the population of any one indicator species. Numerous plant and animal species (e.g., morel mushroom, Bicknell's geranium, jewel beetle, Mountain Bluebird, Black-backed Woodpecker, boreal toad) are nowhere more abundant than in severely burned forests. If we couple basic ecological facts like this with the additional fact that salvage logging

Fig. 13.3 Exactly zero fire-dependent bird species were detected in this salvage logged forest (in precisely the same location where there were plenty of Black-backed Woodpeckers the year before). Is this type of postfire land management that we want to employ across the West?

compromises habitat suitability to those same species then, yes, that is a major problem.

And let's not ignore the effects of cutting trees on the future composition and success of the tree population. Remember my discussion of how seeds are retained in cones on standing dead pine and larch trees for years following severe fire? Fire-killed trees contribute to seeding when mature cones in the burned crown open and disperse seed. Even though the reproductive success of those burned trees is attached to those cones, and the same trees are frequently the target of salvage logging operations. Nothing is more at odds with the needs of the most fire-dependent plant and animal species than salvage logging after fire…nothing. If there were ever an example of how we are failing to manage our public lands in a way that maintains the ecological integrity of the same lands, salvage logging is it.

Frequently, there are additional postfire management activities that are tightly coupled with postfire salvage logging. These activities could use more thought and more robust monitoring for potential negative ecological effects on organisms that benefit from infrequent severe-fire disturbance. One involves seeding, mulching, and afforestation activity designed to reduce the effects of soil runoff following fire. We're told by Burned Area Emergency Rehabilitation (BAER) teams that a burned mountainside almost void of vegetation cannot hold back or absorb water because torched hillsides can turn hydrophobic, or water repellant. Land managers want to prevent whole hillsides from turning to liquid and gushing downslope, but, again, there are all kinds of ecological benefits to stream systems from a disturbance event like that. It's one thing to prevent floods from threatening road integrity or water supplies or communities downstream, but quite another to prevent the natural periodic restructuring of stream systems outside those areas—disturbance is needed by aquatic insect and fish and plant communities alike. It's hard enough for terrestrial ecologists to convince the public that change is natural and that there are ecological benefits following a timber-destroying, stasis-upsetting, catastrophic wildfire, but aquatic biologists face an even greater challenge. Most people probably feel that soil runoff following severe fire is *unnatural* because it unleashes mudslides and fills streams with sediments, which damages (in the short term) the habitat of endangered varieties of salmon and trout. Precisely the opposite is true. Debris-laden flood events following severe fire are perfectly natural disturbance events that serve to stimulate stream channel reorganization, recruitment of large woody debris, pool development, and sediment recharge, which are real benefits to fisheries over the long haul [21].

Tree planting after fire is another common but vastly understudied (but almost certainly ecologically negative) post-fire management activity. I mention that activity here because it often goes hand in hand with salvage logging. There can be no better illustration of the failure to honor a commitment to the National Forest Management Act than arguing in a scoping document that because there is an "…existing allocation of the majority of the area to timber production in the Forest Plan, there is a need for tree planting." While it seems perfectly fine to designate specific stands as part of the timber base, that does not mean we should re-engineer successional pathways after fire to speed tree recruitment and growth for the single purpose of maximizing timber production. By planting after virtually every single fire, we are necessarily converting an ever-growing number of forest patches into little more than unnaturally stocked tree farms. How can the species that occur only during early successional stages possibly benefit from management designed to cut short those early stages (sometimes even including herbicide treatment to eliminate pesky shrubs) to race toward an artificially engineered mature-forest condition as rapidly as possible?

Some foresters argue that planting is needed because tree recruitment from nearby green-tree edges will sometimes be insufficient to allow forests to "recover quickly" after a severe fire, but what's wrong with that? Tree planting cuts short the time needed by early successional plants and animals to have their day before reforestation takes place. Tree planting also reflects a failure to appreciate the role of birds and the role of standing dead trees (which retain cones on their uppermost branches) in the reforestation process. Unless we begin to recognize the fact that disturbance is a natural and necessary and sometimes slow process, and that early seral forest conditions deserve their time in the sun [22], we'll continue down this path of ecologically uninformed land management. There is more to a forest than the trees, and salvage logging and tree planting represent an approach to management that is little more than tree farming. I find it unsettling that so many of us have been duped into believing that a burned forest is a destroyed forest and that the only value remaining is that which can be "salvaged" from the wood, and that widespread tree planting in our national forests is always a wonderful thing.

If land managers were more strongly versed in disturbance ecology, they would understand that a newly burned forest just experienced "recovery" from the perspective of plant and animal species needing early-successional, burned-forest conditions. The intent of laws like the National Environmental Policy Act and National Forest Management Act is to ensure that public lands provide multiple uses on their own terms, meaning that we should certainly be allowed to extract timber, recreate, fish, hunt, and everything else, but if and only if we do these things at a level or intensity that the forests provide on their own terms. Thus, the tree species cover, degree of stocking, rate of growth, and stand conditions should be dictated by the forest, not by the timber shop. If we don't focus on the multiple benefits our public lands provide, then we are treating those lands no differently than timber companies treat their private timberlands.

Perhaps the most common justification (written in USFS scoping documents) for salvage logging is that such logging is an "opportunity." Salvage logging is an opportunity only because environmental regulations are removed and because line officers (with public support) choose to recognize no value other than the commercial value associated with timber harvesting. The boilerplate language in most USFS proposals to salvage log after fire state that the goal is to "...recover merchantable wood fiber affected by the (Put Name in Here) Fire in a timely manner to provide forest products to the local timber industry, contributing to short term timber supply and long-term sustainability of timber on National Forest System lands." This kind of commodity-driven goal is not only short-sighted, but also inappropriate when it so clearly conflicts with the more important and legally mandated goal of sustaining ecological integrity in our forest systems.

There are plenty of green-tree forest stands we can harvest in a sustainable fashion, so claiming that we must use burned timber as an "opportunity" to harvest timber is a weak justification at best. Still, every single time the USFS proposes to salvage log burned timber, district rangers use additional justifications that simply do not hold water. Rangers state that they plan to harvest only a tiny fraction of what's there anyway, so there will still be plenty of dead trees left out there. However, the most fire-dependent plant and animal species occur almost entirely within that small part of the burn slated for salvage. All burned areas are not equal. The clear failure to understand this basic fact becomes apparent every time a district ranger says in an Environmental Assessment that "…less than two percent of the acres of wildfire that burned on the (Name that) National Forest within the last decade have been salvaged." Even though 2% is a small percentage of the area contained within a mapped fire perimeter, that 2% includes perhaps 98% of the burned forest lands that are suitable to the species that most need burned forest! We modeled the best woodpecker habitat based on tree type and size and density variables derived from published statistics and then superimposed a map of those locations onto a map of the planned salvage logging locations within the 2007 Jocko Lakes fire near Seeley Lake, and they fit like a glove. Others have published habitat suitability models for Black-backed Woodpeckers [23], so you can produce your own map the next time a fire burns in your neck of the woods to see this problem for yourself. Any land manager who claims that any negative ecological effects would be insignificant given the small proportion of burned timber salvaged is avoiding the issue at best and practicing public deception at its worst.

Coupled with the "opportunity" and the "we're only taking a little" justifications for salvage logging, there is always what every district ranger uses as a piggyback justification for salvage logging—a pressing need to "help the local economy." That sounds good, but if severe fire creates a once-in-a-lifetime opportunity for people to experience the magic I described earlier in this book, then why are district rangers so fixated on the belief that salvage logging is the only way to help the local economy? There's more than one way to capitalize on the economic opportunity offered by a burned forest. For those who have never experienced the

Fig. 13.4   Communities situated near severe-fire-dependent forests could take advantage of the education and eco-tourism potential associated with the magical conditions that appear following fire. Salvage logging is not the only way to stimulate the local economy following a nearby fire

wonder of a burned forest, severely burned forests are nothing less than magical biological gems (Fig. 13.4). Salvage logging after fire is seen as an "opportunity" only by those who fail to understand that an even better "opportunity" lies with celebrating the ecological uniqueness of conditions severe fire creates. There are plenty of additional places in the timber base where we can harvest green trees with a much lower ecological cost if we find ourselves in urgent need to provide a special boost to the local economy through logging.

References

1. Hutto RL (2008) The ecological importance of severe wildfires: some like it hot. Ecol Appl 18:1827–1834
2. Lloyd H (1938) Forest fire and wildlife. J For 36:1051–1054
3. Sabalow R (2021) Why top California fire expert is so worried about Caldor Fire as it approaches Tahoe. Sacramento Bee
4. Sabalow R, Kasler D (2021) 'Self-serving garbage.' Wildfire experts escalate fight over saving California forests. Sacramento Bee; Thompson, D. 2021. Impact of forest thinning on wildfires creates divisions. AP News. Associated Press
5. Reilly MJ, Zuspan A, Halofsky JS, Raymond C, McEvoy A, Dye AW, Donato DC, Kim JB, Potter BE, Walker N, Davis RJ, Dunn CJ, Bell DM, Gregory MJ, Johnston JD, Harvey BJ, Halofsky JE, Kerns BK (2022) Cascadia burning: the historic, but not historically unprecedented, 2020 wildfires in the Pacific Northwest, USA. Ecosphere 13:e4070
6. Syphard AD, Keeley JE, Gough M, Lazarz M, Rogan J (2022) What makes wildfires destructive in California? Fire 5:133
7. https://www.elementalfilm.com/. Accessed 1 July 2025
8. Hutto RL, Flesch AD, Fylling MA (2014) A bird's-eye view of forest restoration: do changes reflect success? For Ecol Manag 327:1–9

9. Urza AK, Hanberry BB, Jain TB (2023) Landscape-scale fuel treatment effectiveness: lessons learned from wildland fire case studies in forests of the western United States and Great Lakes region. Fire Ecol 19:1

10. Ray C, Cluck DR, Wilkerson RL, Siegel RB, White AM, Tarbill GL, Sawyer SC, Howell CA (2019) Patterns of wood-boring beetle activity following fires and bark beetle outbreaks in montane forests of California, USA. Fire Ecol 15:21

11. Keeley JE (2023) Spatial and temporal strategies of resprouting and seeding in a chaparral shrub species. Ecology 104:e3984

12. Hoecker TJ, Turner MG (2022) A short-interval reburn catalyzes departures from historical structure and composition in a mesic mixed-conifer forest. For Ecol Manag 504:119814

13. Levy S (2022) The return of intentional Forest fires: scientists look to indigenous practices. Bioscience 72:324–330

14. Halsey RW, Syphard AD (2015) High-severity fire in chaparral: cognitive dissonance in the Shrublands. In: DellaSala DA, Hanson CT (eds) The ecological importance of mixed-severity fires. Elsevier, pp 177–209

15. Cohen J, Strohmaier D (2020) Community destruction during extreme wildfires is a home ignition problem. Wildfire Today

16. Higuera PE, Cook MC, Balch JK, Stavros EN, Mahood AL, St LA, Denis. (2023) Shifting social-ecological fire regimes explain increasing structure loss from Western wildfires. PNAS Nexus. https://doi.org/10.1093/pnasnexus/pgad005

17. Wigglesworth A (2023) Legal battle over fire tactics. Los Angeles Times.

18. Johnson EA, Miyanishi K, Bridge SRJ (2001) Wildfire regime in the boreal forest and the idea of suppression and fuel buildup. Conserv Biol 15:1554–1557

19. Hutto RL (1995) Composition of bird communities following stand-replacement fires in northern Rocky Mountain (U.S.A.) conifer forests. Conserv Biol 9:1041–1058

20. Hutto RL (2006) Toward meaningful snag-management guidelines for postfire salvage logging in North American conifer forests. Conserv Biol 20:984–993
21. Jackson BK, Sullivan SMP, Baxter CV, Malison RL (2015) Stream-riparian ecosystems and mixed- and high-severity fire. In: DellaSala DA, Hanson CT (eds) The ecological importance of mixed-severity fires: nature's phoenix. Elsevier, Amsterdam, pp 118–148
22. DellaSala DA, Bond ML, Hanson CT, Hutto RL, Odion DC (2014) Complex early seral forests of the Sierra Nevada: what are they and how can they be managed for ecological integrity? Nat Areas J 34:310–324
23. Latif QS, Saab VA, Haas JR, Dudley JG (2018) FIRE-BIRD: a GIS-based toolset for applying habitat suitability models to inform land management planning. U.S. Department of Agriculture, Forest Service, Rocky Mountain Research Station, Fort Collins

14

Interlude III

If little remains after a severe fire except for a bunch of blackened trees, where do all the birds I've been writing about nest? I mentioned the ground-nesting habits of some species, but nearly a third of the 60-plus bird species that I have found nesting successfully in a recently burned forest use the standing dead trees. I'm sure you've seen a robin nest made of grasses and mud situated in some shrub or green tree around your house, but where in the heck does an American Robin nest in a burned-up forest with nothing left but standing dead trees? They nest in nooks and crannies and broken tops associated with those standing dead trees (Fig. 14.1)! Every darn cavity-nesting species that uses burned forests either digs or uses already excavated cavities in those blackened snags, and even the cup-nesting Western Wood-Pewee places its nest right in the open, on top of a blackened tree branch (Fig. 14.2).

Mentioning wood-pewees reminds me of another quirk associated with burned mixed-conifer forests—they often

R. L. Hutto, *A Beautifully Burned Forest*,
https://doi.org/10.1007/978-3-032-03180-8_14

Fig. 14.1   Broken-top snags that are present *before* a fire sweeps through a forest become important nest sites for a variety of bird species *after* a fire burns through. Here, an American Robin nest is situated inside a burned-out snag

attract species that one would normally associate almost entirely with cottonwood bottomlands. The pewee has a long, distinctive, buzzy "peer" kind of song that rings out and makes you wonder if you've been suddenly transported to somewhere near a cottonwood tree in the bottomland down by the river. Other bird species that one would normally associate with cottonwood bottomlands include the Common Nighthawk, Lewis's Woodpecker, White-breasted Nuthatch, Tree Swallow, House Wren, and even rarely in summer (but commonly in winter), the Downy Woodpecker. Patterns like these have me asking myself again and again, "What's up with that?" Old cottonwood forests share important structural elements (i.e., large, dead, pealy-barked, broken-branched trees) with burned conifer forests, and that may have a lot to do with attracting the bird species they share.

By the time a decade or so passes after fire, many of the standing dead trees will have blown down or snapped in

Fig. 14.2 A Western Wood-Pewee nest positioned atop a horizontal branch in a beautifully burned forest beneath the Grand Teton

half, but those trees only serve to create more and more nesting opportunities for birds like the House Wren and the Lewis's Woodpecker. New feeding opportunities for White-breasted Nuthatches also emerge from beneath tree bark that begins to peel away on aging trees, yielding lots of food for the relatively large nuthatch. Williamson's Sapsucker also becomes common when it can feed its young bundles of ants that become common in the decaying wood (Fig. 14.3). The populations of many of the species that occupy a burned site during the first few years following severe fire begin to decline noticeably after 6 years, while the populations of these additional species are just reaching their highest levels. Very few of the special bird species I observed in the Pattee Canyon burn area can be seen there now, 45 years later, and they won't occur in that place again until the next time the forest burns some 100 years or more into the future! Meanwhile, many of them now occur primarily in other places that have been allowed to burn

Fig. 14.3 Williamson's Sapsucker brings ants to its cavity-dwelling nestling. The bird is rare in the first few years following fire, but common a decade later. These time lag effects illustrate the difficulty of uncovering the positive effects of severe fire through studies that take place only relatively soon after fire

severely enough to create those special postfire conditions. Remember what I said about the importance of maintaining a "shifting mosaic" of naturally occurring postfire forest stands?

I should also mention the shrub-nesting birds that colonize burned areas a decade or two following a severe forest fire. Most every shrub species in Montana (and throughout the West) is capable of resprouting after fire, so birds that use shrubs for nesting or feeding purposes become common as those shrubs begin to flourish. Resprouting shrubs

Fig. 14.4   The Lazuli Bunting's color is enhanced against the back-drop of a blackened conifer forest

bring MacGillivray's Warblers and Orange-crowned Warblers and Lazuli Buntings to name a few and, wow, do those bird species add color to an otherwise blackened land-scape (Fig. 14.4). I already referred to the explosion in numbers of Calliope Hummingbirds once the shrub layer becomes well developed.

The Black-backed Woodpecker's story is where this all started, but I hope I've been able to impress upon you that many other stories of severe-fire dependency exist. I already mentioned the fire morel mushroom and Bicknell's gera-nium, but other plant species that thrive after, or even de-pend on, severely burned conifer forests seem to be endless in number. The populations of some small mammal species explode soon after fire, and furbearers such as the Canada lynx is relatively restricted to mid-succession stages following severe fire [1]. The Colorado firemoth lays eggs on, and is colored to blend in with, fire-dependent blanketflowers [2]. And it's not just one or two beetle species representing insects that have evolved to depend on fire. There are more than 200 pyrophilic insect species whose presence and success on planet Earth depends on fire severe enough and large enough to initiate forest succession [3]. Native boreal toads (*Bufo boreas*) are known to move into severely burned forests in numbers so great that, after one fire in Glacier National Park, rangers closed the Inside North Fork Road because of too much risk to the toads from auto traffic! [4] The list goes on and on and is what I am talking about when I say burned forests are magical.

References

1. Fisher JT, Wilkinson L (2005) The response of mammals to forest fire and timber harvest in the North American boreal forest. Mammal Rev 35:51–81
2. Byers BA (2017) Colorado fires and firemoths. Front Ecol Environ 15:51–52
3. Bell AJ (2023) Like moths to a flame: a review of what we know about pyrophilic insects. For Ecol Manag 528:120629
4. Jamison M (2002) Closed road gives Glacier toads a hopping chance. Lee Enterprises, Missoulian

15

The Philosophy of Land Management Revisited

For any land management agency in the forested West, the overarching management goal should be to allow for multiple use and sustained yield while maintaining the ecological integrity of the same fire-dependent western forest systems. That is the essence of ecosystem management—allow use of the land, but never ever if that use compromises the ability of the land to sustain itself. That is also the law, as outlined in the National Forest Management Act, National Environmental Policy Act, and other ecologically informed legislation. Fortunately, ecologically based land management does not mean eliminating local-community benefits associated with public land management; it means changing the way in which benefits accrue.

If we humans decide that the primary value of a forest is associated with keeping them as green as possible while we harvest the wood pulp tied up in the trees, then there is plenty of great forest and fire science that can be used to show us how to best grow trees rapidly and protect them

R. L. Hutto, *A Beautifully Burned Forest*,
https://doi.org/10.1007/978-3-032-03180-8_15

from fire. That is undoubtedly the best available *forest* science district rangers refer to when they submit justifications for timber harvests. If, on the other hand, we begin to recognize that public lands represent more than tree farms because they supply other values that are far greater, like clean air and water, interesting animals and plants, interesting smells, colors, sounds, and other components of aesthetic beauty, then we can also begin to use the best available *ecological* science as a foundation for understanding how to maintain those greater values while still allowing multiple use. Management designed to promote one condition (mature, green-tree forests) over another (burned forests) is not only futile, but intellectually and (if we are serious about maintaining the ecological integrity of the systems we manage) morally corrupt. Ecological science must be given more weight in leadership positions and elsewhere so that severe fire and the resulting conditions can receive the positive management attention they deserve.

To know whether we are maintaining disturbance-dependent systems that sustain us all while still meeting the needs of humankind without excessive management interference we must (1) understand the evolutionary history of that system, which requires a good understanding of the historical range of natural variation related to fire, and (2) monitor the ecological effects of our land management practices [1]. The environmental conditions within which extant species evolved is what we should be maintaining if we don't want to risk losing it all (ourselves included). That would include actively managing for the maintenance rather than the exclusion or mitigation of natural severe wildfire disturbance events in our forests. Ask your local district ranger, "How are you allowing for the continuation of severe fire disturbance while educating the public about how to stay safe in the face of these natural and necessary forest restoration events?"

We need a paradigm shift in land management agencies from a focus on maintaining static conditions to a focus on supporting sustainable disturbance dynamics [2]. That paradigm shift should include sustaining not just low-severity, but also high-severity, fire disturbance dynamics. We can't emulate fire effects through mechanical means. Thus, any timber harvesting conducted on public lands should fit well within broader landscape plans that are built specifically to maintain the integrity of a dynamic, disturbance-dependent forest system.

The only way to manage mixed-conifer forests for the maintenance of severe-fire disturbance events and the resulting conditions associated with those events is to encourage ecologically meaningful landscape-scale management. That does not mean applying some kind of favored treatment at the landscape scale; it means maintaining postfire patches across the entire forest landscape to ensure there are always patches of every distinct stage of forest succession at any one point in time. This can be accomplished by mapping and maintaining an historically relevant number of large, undisturbed, severely burned forest patches that are, as an example, less than 5 years old. Burned-forest patches that were heavily managed before the fire couldn't be included in the count because they cannot support many of the species that occur only in an *undisturbed* forest that subsequently burns. In general, forest managers ought to be maintaining historically relevant numbers and total areas consisting of a "shifting mosaic" of patches of *all* postfire age categories, including old growth. If they're not, then they are failing to manage for the ecological integrity of the lands that sustain us all. Complex, early-seral forest conditions created by severe fire are especially rare and biologically unique, and they deserve more positive management attention than they currently receive [3]. Only severe fire can create natural successional pathways that will be used by fire-dependent species.

As I already noted, the main reason land management agencies do not celebrate the presence of severe fire in our forests is what appears to be a lack of awareness of the biological significance of severely burned forests. Beyond that, the reason managers ignore the needs of disturbed-forest-dependent species when proposing the most ecologically problematic logging practices is a commodity-driven one—severe fire removes merchantable timber, which is, in their view, an unacceptable loss. Sometimes we're unable to see the forest for the trees. The National Forest Management Act was supposed to change all that when, in 1976, Senator Hubert Humphrey noted that "The days have ended when the forest may be viewed only as trees and trees viewed only as timber. The soil and the water, the grasses and the shrubs, the fish and the wildlife, and the beauty of the forest must become integral parts of the resource manager's thinking and actions." I hope the fire story I have presented here can help revive that sentiment.

Land managers need to publicly acknowledge the ecologically special nature of severely burned forests in much the same way they acknowledge the special nature of old-growth forests—both vegetation conditions are rare and sensitive enough that they should receive minimal management interference. Some places are in and of themselves too special to be altered by logging operations—an old-growth forest is one and a severely burned forest is another. How can we all help ensure that ecosystem management remains the umbrella under which all management activity occurs? We can use voting and public participation opportunities to influence public land management decisions.

References

1. Hutto RL, Belote RT (2013) Distinguishing four types of monitoring based on the questions they address. For Ecol Manage 289:183–189
2. Jones GM, Thompson C, Sawyer SC, Norman KE, Parks SA, Hayes TM, Hankins DL (2025) Conserving landscape dynamics, not just landscapes. BioScience 75:409–415
3. Swanson ME, Franklin JF, Beschta RL, Crisafulli CM, DellaSala DA, Hutto RL, Lindenmayer DB, Swanson FJ (2011) The forgotten stage of forest succession: early-successional ecosystems on forest sites. Front Ecol Environ 9:117–125

16

Toward a Solution

The good news is that if public land managers are willing to modify the way they conduct pre-fire restoration, firefighting, and postfire salvage logging activities, they can maintain both the ecological integrity and multiple use of those disturbance-dependent mixed-conifer forests. I believe several tasks will be required if we are to move closer to a more ecologically enlightened form of land management—one that would minimize interference with the creation and maintenance of the special burned-forest conditions I have highlighted in this book.

Educate the Public About the Need for Ecosystem Management

Before we can eliminate the threats associated with land management activities conducted before, during, and after fire, we first need to ensure that all people come to

R. L. Hutto, *A Beautifully Burned Forest*,
https://doi.org/10.1007/978-3-032-03180-8_16

177

understand the ecological value of maintaining natural severe fire disturbance in western mixed-conifer forests. The special nature of severely burned forests is currently cloaked in secrecy while severe fire itself is demonized. It's tough for people to begin loving what they have been taught to hate, but if we want to let evolutionarily important fires (yes, even severe, stand-replacement fires) into most conifer forest systems, attitudes must first change. That, in turn, starts with a *change in our public messaging* to highlight the ecological value associated with disturbed-forest conditions. The task of educating the public-at-large, politicians, land managers, and other decision makers about the need to manage for the creation and maintenance of disturbed-forest conditions and associated early successional species will be a difficult task. That task is summarized well by Alan Taylor who wrote 50 years ago that "…the greater challenge facing man is to gain and diligently apply new biological insight concerning the roles of lightning and fire in plant and animal communities, to couple this knowledge with technological capabilities to prevent disasters associated with lightning, and yet to allow this change agent to pursue, to a considerable extent, its natural course" [1].

A quick look at some of the public messaging surrounding fire shows just how bad the current messages really are. They do not reflect any kind of attempt to celebrate the process of fire disturbance and the resulting burned-forest wonderland I described in previous chapters. Instead, fire-related messages are designed to help achieve utilitarian management goals by seeking to bolster public support for the same problematic land management activities I focused on earlier. People in power still consider fire to be the enemy, as evidenced by the militaristic language surrounding descriptions of severe fire. Just look at the words Montana governor, Judy Martz, used during her state of the state address in 2001 following the fires of 2000 in Montana—she said, "We *survived* one of the worst *natural disasters* in the history of Montana…Never before on American soil has such a *large army fought a single enemy*"! That kind of language (and the message it carries) does not encourage us to embrace fire and view it in the same way we view the sun and rain as natural and necessary but potentially dangerous physical entities. We need to learn to celebrate what fire brings while learning to live safely in its midst, and that will require political leadership.

Take a minute to think about it. Most of us were inundated with Smokey Bear's messages beginning just after we were weaned from the baby bottle. The "True Story of Smokey Bear" comic book paints an entirely negative view of fire and ends with Smokey telling kids "…how he hates forest fires," and "…how they destroy his animal friends!" Among the most telling of Smokey's posters was one printed in 1951 where Smokey stands pointing to a huge forest fire behind him with the following headline emblazoned across the top: "This shameful waste weakens America!" Technically, Smokey is referring to human-caused fire with his "Only you can prevent forest fires" slogan, but the overarching message that comes across in the comics and posters is that wildfire (severe crown fire in particular) is a universally bad outcome, human caused or not. Even as recently as 2014, the Smokey Bear web page read: "Wildfire is one of the most destructive natural forces known to mankind." Why not one of the most creative forces known to mankind? The "all fires are bad" message has now morphed into one no less troublesome. The new message now emerging out of the mouths of public land management agencies, some conservation organizations, and even many fire scientists is that low-severity fire or prescribed fire is *good*, and overstory crown fires are *bad*. This is what I call the good fire/bad fire message; it acknowledges that there is such a thing as good fire, but it's clearly not fire with big flames! To this point in time, we've allowed low-severity understory fire to pass through the door of acceptability, but high-severity fires are still viewed by most as an unnatural result of fuel accumulation from decades of fire suppression [2].

Even worse than not accepting severe fire as a natural disturbance process, there is some current positive messaging that includes the terms "fire-adapted communities" or "fire-dependent forests," but only in reference to low-severity fire, which is misleading. As I have tried to make abundantly clear in this book, the most dramatic fire adaptations one can point to and the most definitive examples of fire dependency are the result of evolution in the face of high-severity, not low-severity fire. Smokey Bear carried no nuance to his message, and the recent "good fire/bad fire" message carries no nuance either. Meaningful messaging would reflect the fact that low-severity understory fire regimes apply to a small proportion of forest area across the West and that, in most forest types, overstory fires are good and are not threatening if we learn to be "Firesafe." We ought to be talking about fire in a manner more akin to the way we talk and message about other natural phenomena that are inherently dangerous but tolerable if we're prepared i.e., use sunblock when outdoors to protect yourself from harmful effects of the sun, or carry pepper spray when in grizzly country to protect yourself against a charging bear.

The most troublesome aspect of the current "good fire/bad fire" message is that, even in the scientific literature, severe fires are depicted as unnatural and bad, and they occur only because our forests are out of whack due to management errors in the past. One can hear about the problem with our forests through every radio, television, newspaper, magazine, and social media outlet that exists. I am surprised that organizations like the Nature Conservancy—organizations supposedly devoted to conservation—keep pushing this "good fire/bad fire" story in every fire-related publication they produce [3]. One would think that a conservation organization would be sensitive to the needs of plants and animals in *all* fire-dependent systems, which means educating the public about the nuanced differences in fire regimes and helping them understand that most western mixed-conifer forests need healthy doses of severe fire.

As I noted earlier, most fire scientists would agree that a small percentage of low-elevation, dry forest types in the western US support a frequent, low-severity fire regime. Most fire scientists would also agree that high-elevation lodgepole and spruce-fir forests support a severe crown-fire regime. The middle ground, however, is where most conifer forests and their mixed-severity fire regimes lie. It's encouraging to see fire scientists acknowledge that there are marked differences in fire regimes between the high- and low-elevation conifer forest types, but more nuanced regime changes are largely ignored when it comes to that middle ground. Instead, land managers seem to view and manage the middle ground as if those forests are typified by a low-severity, frequent-fire regime, so we spend billions on thinning and prescribed burning to moderate fire behavior in those forests even though severe fire is an important natural component. Mixed-conifer forests are ecologically unique and deserve management that embraces healthy doses of severe fire.

The basic problem is that the "forests are out of whack" story has been extended to the broadly distributed mixed-conifer-forest middle ground where it doesn't belong. Indeed, all the public ever hears in reference to the mixed-conifer forest zone, is that we have an unnatural fuel "problem" that needs to be addressed through pre-fire management (thinning, prescribed fire). The "forests are out of whack" problem is way overblown. The plethora of severe-fire adaptations expressed by the plants and animals that live in western mixed-conifer forests provide strong evidence that most of those forests were born of, and are now maintained by, healthy doses of severe fire, not low-severity fire. It's time to give the severe-fire story more airtime and for the messaging surrounding fire to change toward something that reflects the nuance by discussing why severe fire is, in most forests, operating as it always has and that this is a good thing. Maybe there's room for a "let's not get too carried away; severe fire still plays a critical role here" story. Only then will our western mixed-conifer forests begin to be managed in a manner designed to maintain severe fire disturbance.

In summary, the path toward accepting severe fire as a natural and necessary part of mixed-conifer forest ecology will require changes in public messaging through pressure from all of us and that will, in turn, require abandoning the current view that fire is an enemy. We need to start with education about the wonderful nature of, and ecological importance associated with, severe crown fire in our fire-dependent conifer forest systems. Before that, we need to stop inundating the public with warlike language when it comes to severe fire and with images that play on a "good fire/bad fire" theme designed to spread the falsehood that any burned-forest condition other than one following thinning and prescribed burning is undesirable. I love to see people get excited about forest regrowth or "recovery" after fire, but even that sentiment misses the point entirely. Evidence that the forest is greening up and is on its way to restoring what was lost may make us want to celebrate, but that focus takes all the attention from an even *more* beautiful story—the magical forest transformation that already occurred and will persist only within the first few years following fire. The fleeting gift of a burned forest itself will always be overlooked if one's focus is set entirely on regaining a green forest just like the one that was there immediately before fire. Don't be one of those people who have been misled into thinking that a green, unburned forest is the only forest condition we should strive to maintain on our public forestlands! Once we break from a green-tree mindset, we can then start delivering ecologically informed messages that celebrate and embrace severe fire as an important component of our western conifer forests. With ecological education, the public can come to view a newly blackened hillside not as a scar, but as a beauty mark on the landscape (Fig. 16.1)!

Reconsider Pre-fire Management Plans for Mixed-conifer Forests of the West

When it comes to the plants and animals that depend on severely burned forest conditions, there is a major problem associated with pre-fire thinning and prescribed burning in our mixed-conifer forests. As outlined above, a forest that is altered *before* fire becomes less suitable to fire-dependent species *after* fire. That means the scale and spatial extent of pre-fire logging must be limited enough to ensure there will always be a constant supply of relatively unaltered mature forest available for severe fire to periodically initiate the process of natural succession and the creation of those burned-forest conditions that many species depend on.

What could we be doing differently with respect to pre-fire thinning and prescribed burning? The first thing we

Fig. 16.1 Most of us have been groomed to believe that the aftermath of a severe forest fire (such as the Black Mountain fire pictured here) is a scar on the landscape. Instead, the burned forest creates a beauty mark on the landscape

could do is dispel the myth that most every forest needs pre-fire thinning or prescribed burning to "restore" forest structure and composition or to reduce fire severity. Except for a relatively small proportion of western conifer forest acreage that sits well within the lowest-elevation and driest forest types, there is no need for some kind of tree-thinning restoration; most conifer forest acreage is well within historical range of natural variation in terms of tree composition and density and is perfectly fine as is. Thinning and burning in the name of restoration in most conifer forest types is, therefore, unjustified.

With greater home safety in place, perhaps we could move toward allowing more naturally occurring Wildland Fire Use. The desire by land managers to experiment with re-introducing mixed-severity fire through landscape-level fuel treatments conducted at the right time of year is a promising recent development. I am elated to see the acceptance of severe fire as an important ecological component that some forest managers are trying to maintain on the landscape (in small doses so far, but it's a step in the right direction). Meanwhile, we have a fire-industrial complex that continues to sell the idea that we have a fire problem that can be addressed only through massive vegetation treatments and prescribed burning. Should we, the public, support these efforts to alter the natural fire disturbance process? Just listen to what all the fire-dependent plants and animals have to say; the answer is clear.

Does all this mean we can't cut trees anywhere in our state and national forests? Absolutely not. There is plenty of room to cut trees in an ecologically sustainable fashion while leaving most mature and older forests alone. By leaving most of our mature forest areas alone, they could eventually be transformed by severe fire into a beautiful succession of *unaltered* forest conditions, each supporting its own unique species complex. A small amount of green-tree harvesting for wood products fits perfectly well within the mission of the USFS and could be ecologically sustainable, but that would also mean abandoning ecologically inappropriate "restoration" justifications for timber harvests in mixed-conifer forests outside the Wildland-Urban Interface. Forest managers could achieve a small measure of safety by creating defensible space by targeting their sustainable timber harvesting toward small areas immediately adjacent to communities. At the same time, forest managers could be much less aggressive and more fiscally prudent by cutting back on forest thinning and prescribed burning beyond the Wildland-Urban Interface. Because safety is unrelated to forest conditions miles away from human settlement areas, safety-based thinning treatments should be focused on providing zones for suppression personnel to use in staging their activities, not on trying (and failing) to affect severe fire behavior across the landscape. To mitigate the risk of structure losses during wildfires, we ought to focus on houses first and move outward from there. That single shift in management emphasis could probably supply our timber needs in perpetuity while making structures safer during less extreme fire events.

Focus Firefighting Efforts on Safety, Not on the Forests Themselves

How do we move beyond our well entrenched but misguided attempts to suppress nearly every fire? We do so by moving toward an ecologically based wildland fire use policy because land management designed to let fires run their course will do the best possible job of periodically restoring forest conditions beyond the Wildland-Urban Interface. Am I suggesting we return to some kind of "let it burn" policy? Yes, pretty much. Importantly, such a policy would not compromise public safety because safety is, as described earlier, a home ignition issue that can be solved through becoming "Firesafe." The combination of home hardening, forest treatments focused on areas close to the Wildland-Urban Interface (the only activities that affect safety), and a policy allowing managed fire for resource benefit would go a long way toward reducing wildland fire management activities that now consume more than half the annual USFS budget. Eliminating firefighting activity in the outback would not sacrifice public safety one bit and would save a large fraction of the billions we spend annually on firefighting. Public lands could then devote a lot more time and money toward things like recreation—a need made abundantly clear during the recent global pandemic.

Don't Destroy the Special Burned-Forest Conditions That Nature Creates

What can we do differently with respect to postfire salvage logging? We can eliminate it. Basically, postfire salvage logging should not be allowed except to address very narrowly focused safety issues following severe fire. Ecologists like me are not at all opposed to timber harvesting; we are opposed to harvesting in the most ecologically sensitive places on the planet, which include federally designated Wilderness areas, old-growth forests and, yes, severely burned forests. Burned forests should be the LAST, not the first place we go to get our wood (in a legislatively expedited fashion designed to waive environmental regulation, no less). There are lots of places where we can get wood with little to no environmental impact—burned forests are not among them. Forests also do not need our "help" through seeding, tree planting, spraying, or any other postfire activity designed to speed up or interfere in any way with the natural succession process.

There is no mixed-conifer forest condition on planet Earth where biological uniqueness and beauty are greater than in the first few years following a severe fire. Thus, the potential negative effect on local economies can be offset by capitalizing on the ecotourism potential associated with the newly transformed forests that harbor once-in-a-lifetime experiences. The wildflower and birdwatching show following severe fire is better than any other nature-based tourist attraction on Earth! Nearby hotels and restaurants could be filled for years following a severe fire if local communities embraced the idea. Tourism was never higher than after the fires of 1988 in Yellowstone National Park. In 1989, a then-record 2.7 million visitors flocked to Yellowstone to learn about the fire. The educational opportunity provided by the Black Mountain fire near Missoula after 2003 allowed thousands of school children and adults to learn about the naturalness and necessity of severe forest fires. As I noted earlier, Missoula may very well be the first community in America to include residents who are informed enough to celebrate the presence of the next nearby fire with a parade downtown...I certainly hope so! We have so much we could do to enhance recreational opportunities associated with burned forests, so why do postfire management plans fail to include economic opportunities outside of timber harvesting?

Don't Let Climate Change Derail Ecologically Informed Land Management

The suspected recent shifts in fire regimes were initially associated with the "forests are out of whack" story. Now, the relatively recent and dramatic climate-related shifts in fire regimes are more commonly associated with a new storyline—we have a "wildfire crisis." The symptoms of the recent crisis are somewhat different from those originally identified by Weaver, Covington, and others, who were driven to explain the relationship between the sudden, turn-of-the-century decrease in fire frequency and subsequent changes in forest structure. The "forests are out of whack" story worked well to explain fuel buildups in the low-elevation dry-forest vegetation types. The current "wildfire crisis" story is something different because it emphasizes climate-based fire regime changes (primarily increases in fire frequency and extent) that have occurred suddenly and pretty much everywhere. One can read most any fire-related story in the popular press (or even in the professional fire-related publications) and see the shift from an emphasis on poor forest management toward a climate-based wildfire crisis. Indeed, in 2023 the Nature Conservancy and the Aspen Institute convened workshops devoted to creating a "Roadmap for Wildfire Resilience," in "…response to the *wildfire crisis* (my emphasis) in the United States." The US Forest Service has also recently developed a "10-Year *Wildfire Crisis* (my emphasis) Strategy" [4].

The shift away from the claim that changes in fire behavior is the result of forest conditions is almost certainly because we are witnessing changes in fire behavior in many non-forested as well as in forested environments. There is growing recognition that factors other than changes in forest structure are contributing to the more recent changes we have begun to witness with respect to wildfire behavior. The primary causal factor related to the recent wildfire crisis is not an increase in forest fuels; it is climate-change-induced "fire weather" [5]. Although climate change received little to no attention during the "forests are out of whack" period, we now recognize that climate change is the single most important factor accounting for changes in fire frequency and severity we're seeing today. The changes in fire behavior we're seeing today are *not* the result of "100 years of mismanagement" in our forests, as some continue to argue [6]. Forest conditions (i.e., fuel loads) account for less than a sliver of the variation in fire size and severity we are witnessing today. Instead, the broadscale increase in wildfire frequency across the western United States has been driven primarily by sensitivity of fire regimes to recent changes in climate over the entire planet.

Within the low-elevation dry forests, climate-influenced megafires are now creating decreases in old-growth forest cover, and postfire successional pathways are more likely than ever to lead to vegetation type conversion [7]. I find it disconcerting, however, to see scientists argue that what we need to do is not to address the root of the problem (climate change, the statistically dominant reason behind the trends), but to embark on large-scale restoration ("managed dynamics" in their words). Reporters continue to parrot the scientists who propose cutting and burning as solutions to a climate problem. A recent front-page story in the San Diego Tribune entitled "Fire-safety effort thinning trees from region's forests" by Joshua Smith (22 January 2023) highlights a forest-thinning project on Palomar Mountain in the Cleveland National Forest where loggers cut down hundreds of white fir and incense cedar on a steep hillside dotted with rustic homes. The cutting was "…aimed at protecting a forest on the front lines of climate change." The very next sentence dismisses the climate issue altogether, however, and goes right on to blame past land management activity for the problem! Smith writes "Experts say a century of fire suppression has allowed thickets of younger trees to crowd old growth stands across the West, increasing the possibility that mega blazes will obliterate drought-stressed ecosystems." We seem to be stuck in a rut.

Recent changes in forest conditions are *primarily* a climate problem and can be solved only by addressing climate change, not by thinning forests, prescribed burning, tree planting, salvage logging, or anything else that fails to strike at the heart of the climate problem. To deal with the current wildfire crisis, the USFS developed a strategy to "…work with partners to engineer a *paradigm shift* in the way we manage public and private lands." A paradigm shift sounds great. However, instead of reading that the USFS plan is to educate the public about the naturalness and necessity of severe fire and about how to be safe in the face of natural fire disturbance events, I read that their paradigm shift involves conducting "…fuels and forest health treatments more strategically and at the scale of the problem." What? Do they really think doing the same thing (thin and burn), but at a much larger spatial scale (a 20-million-ha increase) constitutes a meaningful paradigm shift? I wish they had acknowledged that the only way out of an ever more flammable world is to accept that we have an existential climate problem (one caused by fossil fuel burning and, ultimately, unchecked human population growth) that needs to be addressed. The wildfire crisis is merely a symptom of climate change—excess fuel in our forests, grasslands, and shrublands is not the problem. We cannot stop the growing severity and extent of fires by re-doubling our effort to thin and burn. That's merely diverting attention from the real culprit, which is climate change.

Even though average fire sizes and frequencies may have moved well beyond historical limits, that doesn't mean severe fire itself is bad or unnatural. Ecologically relevant past environments are still relevant targets if we want to keep all the plant and animal species with which we share this planet. Fortunately, severe-fire-dependent mixed-conifer forests do not appear to be as easily compromised by climate change as the low-elevation dry forest types are. In fact, the *proportion* of an area that burns severely is not much different than it has ever been across most western forest types. That proportion has not changed significantly between 1984 and 2017, especially in the more mesic mixed-conifer forest types [8]. When a crown fire burns across a swath of forest, it nearly always results in a beautiful mosaic of severities—just look at any MTBS fire severity map to see for yourself [9].

Global warming trends also do not "change everything" such that thinning and prescribed burning (the first of the three problematic land management practices outlined earlier) somehow become helpful. Those practices still do not affect change in fire behavior during hot, dry, windy conditions and they are still ecologically inappropriate land management activities that should be greatly modified in location and extent. It is impossible to sustain the biological integrity of a forest system in the face of goals set primarily by a narrow focus on tree growing or on reducing the prospect or amount of severe fire in the future. It is quite possible, however, to maintain biological integrity in the face of limited and sustainable timber extraction. We've seen where an entirely utilitarian approach can lead us—toward the embarrassing predicament of having to say that our management efforts (to suppress fires) were badly misguided. I don't think we want to make a similar mistake by now embarking on widespread forest thinning and other pre-fire and postfire "restoration" initiatives that may be equally misguided.

I hope my earlier descriptions of a beautifully burned forest have given you a feel for the gift that only a severe fire can bring. I also hope that the five steps outlined above can be realized. In the meantime, let me offer you a landscape view of the 1988 Scapegoat fire in Montana (Fig. 16.2). If you stare at that picture for a moment, it will help you think about what goes on in a severely burned forest. That fire brought us not only an incredible scene, but it must also have created something close to heaven for a huge number of plants and animals that find a suitable home under no other forest condition, don't you think? We must move beyond our turn of the century view of fire, as summarized by Gifford Pinchot, the first USFS director, when he wrote, "Of all the foes that attack the woodlands of North America, no other is so terrible as fire." Pinchot was dead wrong. If you're lucky enough to live in the West near a mixed-conifer forest where enlightened land managers allow your local forest to burn in the way it has for millennia, and everyone is safe because they have prepared themselves for such a fire, and they leave it alone after the fire, then you will have been given a once-in-a-lifetime opportunity to see some of the most amazing things that nature has to offer right at your doorstep (Fig. 16.3). We can have it all—disturbance-dependent forests and *all* the things beautifully burned forests provide, but not until…not until we learn to love what we've been taught to hate (Fig. 16.4)!

Fig. 16.2 A strikingly beautiful view of a severely burned forest adjacent to the Scapegoat Wilderness in western Montana

Fig. 16.3 Great Gray Owl shedding light on what makes a severely burned forest so special

Fig. 16.4 Couple of young guys enjoying their walk through a severely burned forest wonderland

References

1. Taylor AR (1974) Ecological aspects of lightning in forests. Tall Timbers Research Station, pp 455–482
2. Hanson CT (2021) Smokescreen: debunking wildfire myths to save our forests and our climate. Kindle edition. The University Press of Kentucky, p 28
3. https://www.nature.org/en-us/what-we-do/our-priorities/protect-water-and-land/land-and-water-stories/why-we-work-with-fire/. Accessed 1 July 2025

4. https://www.fs.usda.gov/managing-land/wildfire-crisis. Accessed 1 July 2025
5. Vaillant J (2023) Fire weather: a true story from a hotter world. Alfred A. Knopf, New York
6. Cornwall W (2021) Clearing the tinderbox. Science 373:1300–1303
7. Steel ZL, Jones GM, Collins BM, Green R, Koltunov A, Purcell KL, Sawyer SC, Slaton MR, Stephens SL, Stine P, Thompson C (2023) Mega-disturbances cause rapid decline of mature conifer forest habitat in California. Ecol Appl 33:e2763
8. Williams JN, Safford HD, Enstice N, Steel ZL, Paulson AK (2023) High-severity burned area and proportion exceed historic conditions in Sierra Nevada, California, and adjacent ranges. Ecosphere 14:e4397
9. www.mtbs.gov. Accessed 1 July 2025

Epilogue

In the time since I began writing this book, numerous fires have burned in ways that firefighters and fire managers have never seen in their lifetimes. Fires fueled by low humidities, high temperatures, and strong winds have transformed what used to be "normal" wildfires into infernos that have burned thousands of homes to the ground. Vaillant's "Fire Weather" [1] captured this change in a way that really affected me. The world is not the same one it was even as recently as a decade ago. If you think some of us had it hard some 30 years ago trying to convince people that severe fire is natural and necessary, the problem suddenly became a thousand times harder. Yes, fires are burning larger areas these days, but they are not producing ashen moonscapes leaving few standing dead trees! Wildfires are still creating some beautifully burned forests, which we can still celebrate when we're lucky enough to get naturally occurring wildfires on our public lands. At the same time, we can learn to live safely amid disturbance-dependent forest systems while addressing climate change (the root cause of our changing

fire environment) so that we do not have to experience unnaturally common or unnaturally large fires in the future.

Reference

1. Vaillant J (2023) Fire weather. Alfred A. Knopf, New York

GPSR Compliance
The European Union's (EU) General Product Safety Regulation (GPSR) is a set
of rules that requires consumer products to be safe and our obligations to
ensure this.

If you have any concerns about our products, you can contact us on

ProductSafety@springernature.com

In case Publisher is established outside the EU, the EU authorized
representative is:

Springer Nature Customer Service Center GmbH
Europaplatz 3
69115 Heidelberg, Germany